ENERGY POLICIES, POLITICS AND PRICES

NATURAL GAS AND ELECTRICITY

A PRIMER ON ENERGY MARKETS

ENERGY POLICIES, POLITICS AND PRICES

Additional books in this series can be found on Nova's website under the Series tab.

Additional E-books in this series can be found on Nova's website under the E-book tab.

ENERGY SCIENCE, ENGINEERING AND TECHNOLOGY

Additional books in this series can be found on Nova's website under the Series tab.

Additional E-books in this series can be found on Nova's website under the E-book tab.

ENERGY POLICIES, POLITICS AND PRICES

NATURAL GAS AND ELECTRICITY

A PRIMER ON ENERGY MARKETS

LEE HUANG
EDITOR

nova publishers
New York

For permission to use material from this book please contact us:
Telephone 631-231-7269; Fax 631-231-8175
Web Site: http://www.novapublishers.com

Library of Congress Cataloging-in-Publication Data

ISBN: 978-1-62417-712-5

Published by Nova Science Publishers, Inc. † New York

CONTENTS

PREFACE

Natural gas is the second largest primary source of energy consumed in the United States, exceeded only by petroleum. A primary energy source is an energy source that can be consumed directly or converted into something else, like electricity. Roughly a third of the natural gas consumed in the United States goes into power plants for the production of electricity. Electricity, a secondary energy source, results from the conversion of primary fuels such as fossil fuels, uranium or wind, into a flow of electrons used to power modern life. This book explores the workings of the wholesale markets for these two forms of energy, as well as energy-related financial markets.

Chapter 1 – Natural gas and electricity are two forms of energy that are of particular interest to the Federal Energy Regulatory Commission. This primer explores the workings of the wholesale markets for these two forms of energy, as well as energy-related financial markets.

Natural gas is the second largest primary source of energy consumed in the United States, exceeded only by petroleum. A primary energy source is an energy source that can be consumed directly or converted into something else, like electricity. Roughly a third of the natural gas consumed in the United States goes into power plants for the production of electricity.

Electricity, a secondary energy source, results from the conversion of primary fuels such as fossil fuels, uranium or wind, into a flow of electrons used to power modern life.

Natural gas and electric markets involve both physical and financial elements. The physical markets contain the natural resources, infrastructure, institutions and market participants involved in producing natural gas and electricity and delivering it to consumers. They also include the trading of and payment for the physical commodity, natural gas. The financial markets include the buying and selling of financial products derived from physical natural gas and electricity. These financial markets also include market structures and institutions, market participants, products and trading, and have their own drivers of demand and supply. In general, physical and financial markets can be distinguished by the products and by the intentions of the market participants involved. Physical products are those whose contracts involve the physical delivery of natural gas or electricity. Physical market participants are those who are in the market to make or take delivery of the commodity. Financial products do not involve the delivery of gas or electricity; instead, they involve the exchange of money.

Physical markets can be further differentiated by:

- **Location:** regions, nodes, zones or hubs;
- **Timeframes:** hourly, daily, monthly, quarterly or yearly;
- **Types of products:** natural gas molecules or electrons, pipeline or transmission capacity and storage; and
- **Nature of sales:** retail sales involve most sales to end-use customers; wholesale sales involve everything else.

Chapter 2 – Due to the growth in natural gas production, primarily from shale gas, the United States is benefitting from some of the lowest prices for natural gas in the world and faces the question of how to best use this resource.

Different segments of the U.S. economy have different perspectives on the role natural gas can play. Suppliers, which have become the victims of their own production success, are facing low prices that are forecast to remain low. Some companies that have traditionally produced only natural gas have even turned their attention to oil in order to improve their financial situation. Smaller companies are having a difficult time continuing operations and larger companies, including international companies, have bought into many shale gas assets. Prices have remained low even as consumption has increased, in part, because producers have raised production to meet the demand and because companies have improved efficiency and extraction techniques. Some companies, many with large production operations, have applied for permits to export natural gas. This has raised concerns from consumers of natural gas that domestic prices will rise. The debate regarding exports is ongoing.

Industries that consume natural gas have seen input costs drop, and some have heralded low natural gas prices as the impetus for a manufacturing revolution in the United States. Some companies have begun to make major investments to take advantage of the low natural gas prices, particularly in petrochemicals. Other companies are waiting to see if prices will remain low long enough to warrant major investments in new facilities. Meanwhile, the electric power sector has already seen a transition from coal-fired generation to natural gas. Low natural gas prices are also putting pressure on renewable sources of power generation. However, increases in demand will put upward pressure on natural gas prices.

The transportation sector, the one part of the economy vulnerable to foreign energy supplies, is beginning to explore ways to use more natural gas. Transportation makes up less than 1% of U.S. natural gas consumption and would require billions of dollars in investment to increase that share significantly.

All of the change that has taken place so far has occurred despite environmental concerns and regulatory developments at the state and federal level that might curtail production. Natural gas is a fossil fuel that produces various pollutants, some more than other fossil fuels and some less. Methane, the major component of natural gas, is also a potent greenhouse gas when released without burning. Other environmental concerns focus on water use and disposal in hydraulic fracturing to extract natural gas from shale formations.

Over the next five years, many of the issues being debated now may be decided. The industry and market are adapting to the newly found supplies and the concerns associated with them, as well as integrating more natural gas into the economy. There are many evolving issues some of which Congress can influence directly because of statutes and some indirectly. On the demand side, legislation has been introduced regarding exports of liquefied natural gas and alternative fuels for vehicles. There has been other legislation related to environmental regulations of natural gas.

In: Natural Gas and Electricity
Editor: Lee Huang
ISBN: 978-1-62417-712-5

Chapter 1

ENERGY PRIMER: A HANDBOOK OF ENERGY MARKET BASICS*

Federal Energy Regulatory Commission

1. INTRODUCTION

Natural gas and electricity are two forms of energy that are of particular interest to the Federal Energy Regulatory Commission. This primer explores the workings of the wholesale markets for these two forms of energy, as well as energy-related financial markets.

Natural gas is the second largest primary source of energy consumed in the United States, exceeded only by petroleum. A primary energy source is an energy source that can be consumed directly or converted into something else, like electricity. Roughly a third of the natural gas consumed in the United States goes into power plants for the production of electricity.

Electricity, a secondary energy source, results from the conversion of primary fuels such as fossil fuels, uranium or wind, into a flow of electrons used to power modern life.

Natural gas and electric markets involve both physical and financial elements. The physical markets contain the natural resources, infrastructure, institutions and market participants involved in producing natural gas and electricity and delivering it to consumers. They also include the trading of and payment for the physical commodity, natural gas. The financial markets include the buying and selling of financial products derived from physical natural gas and electricity. These financial markets also include market structures and institutions, market participants, products and trading, and have their own drivers of demand and supply. In general, physical and financial markets can be distinguished by the products and by the intentions of the market participants involved. Physical products are those whose contracts involve the physical delivery of natural gas or electricity. Physical market participants are those who are in the market to make or take delivery of the commodity.

* This is an edited, reformatted and augmented version of A staff report of The Division of Energy Market Oversight, Office of Enforcement, Federal Energy Regulatory Commission, dated July 2012.

Financial products do not involve the delivery of gas or electricity; instead, they involve the exchange of money.

Physical markets can be further differentiated by:

- **Location:** regions, nodes, zones or hubs;
- **Timeframes:** hourly, daily, monthly, quarterly or yearly;
- **Types of products:** natural gas molecules or electrons, pipeline or transmission capacity and storage; and
- **Nature of sales:** retail sales involve most sales to end-use customers; wholesale sales involve everything else.

Physical Fundamentals

Much of the wholesale natural gas and electric industry in the United States trades competitively; some markets are established through administrative processes based on the cost of providing service. In competitive markets, prices are largely driven by the economic concepts of supply and demand. Underlying the supply and demand for natural gas or electricity are physical fundamentals - the physical realities of how markets produce and deliver energy to consumers and how they form prices. These physical fundamentals will be covered in Chapter 2, on natural gas, and Chapter 3, on electricity.

Wholesale natural gas and electric markets differ from other competitive markets, however, in critical ways. While this primer focuses on wholesale markets, demand is ultimately determined at the retail level. Retail use is relatively inelastic in the short-term, although this may be less so with some larger customers. Retail use of gas or electricity exhibits some unique characteristics:

Limited customer storage options: Retail consumers have few options for storing natural gas and electricity. For natural gas, large consumers and entities that sell to retail consumers may be able to store gas, but smaller consumers do not have this option. For electricity, smaller consumers may have batteries, but nothing adequate to ensure refrigeration, for example. Without storage, consumers cannot buy when prices are low and use their stored product when prices rise. This limits consumers' response to changes in prices.

Substitutes: Retail consumers have few substitutes for natural gas or electricity, certainly in the short-term. If natural gas or electric prices go up, consumers cannot quickly switch to a different product. Longer term, they may be able to switch to gas from electricity for heating, or they may be able to insulate or install new windows or take other steps to reduce their consumption of energy. In addition, demand-response programs can provide benefits to those who would reduce their energy needs at certain times; this might include turning off air conditioning during the hottest part of a day in order to help reduce electric load.

Necessity: Unlike most other products, natural gas and electric service are necessities today, and a lack of service can mean customers without heat, the ability to cook or refrigerate food or the ability to run their businesses. Blackouts and other service disruptions create operating problems and hazards as well. Consumers cannot postpone the purchase of electricity or natural gas. They may be able to turn down their thermostats, but cannot eliminate consumption altogether for an extended period of time.

Because consumers have limited ability to reduce demand, supply must match demand instantaneously, in all locations.

For natural gas, this means production, pipelines and storage need to be sized to meet the greatest potential demand, and deliveries need to move up and down to match changes in consumption. Natural gas has underground and above-ground storage options and linepack, which involves raising the pressure in a pipeline to pack more molecules into the same space. Gas flows through a pipeline at about 15 mph, so new supply can take hours or days to reach its destination. That increases the value of market-area storage, which vastly reduces the distance and time needed for gas to reach consumers.

For electricity, storage is more limited, although technology – involving batteries and flywheels – is being developed. Hydroelectric pumped storage is available in a few locations; this involves pumping water to high reservoirs during times of slack electricity demand, then letting the water flow downhill through electricity-generating turbines when demand for power rises. Generating plants, transmission and distribution lines, substations and other equipment must be sized to meet the maximum amount needed by consumers at any time, in all locations. For all practical purposes, electricity use is contemporaneous with electricity generation; the power to run a lightbulb is produced at the moment of illumination.

Natural gas and electric industries are capital intensive, requiring access to financial markets to support daily operations, trading and investment programs. Access to financial markets requires maintaining an investment grade credit rating to support activities ranging from daily transactions to long-term development of infrastructure.

Financial Markets and Trading

Financial markets are where companies and individuals go if they need to raise or invest money. They are important to natural gas and electric markets in two key ways. First, they provide access to the capital needed for operations. Second, some natural gas- or electricity-related products may trade in commodity markets or, as derivative products (see below), in financial markets.

Natural gas and electricity are traded like commodities, just like metals, corn, wheat or oil. They may not be visible, but you can turn them on and off, and measure them. Commodity markets began as ways for farmers to sell their products, or even a portion of their production before it was harvested, providing them with capital to continue operations.

Commodity markets evolved to provide other tools for farmers (and other commodity producers) to manage their risk, notably the risk of adverse changes in price. These financial products were derived from the physical natural gas and electric products, and are known as derivatives. Since their inception, trading in physical commodities and derivatives has attracted others to the market, such as speculators hoping to make a profit from changes in price.

The market for natural gas derivatives has grown enormously within the past decade, as competitive natural gas and electric markets matured and investors came to see energy commodities as investments, not just a source of power. This trading affects the physical markets in a number of ways, and is discussed in Chapter 4, Financial Markets and Trading.

Market Manipulation

Where there are markets, there will be those who attempt to manipulate the markets for their own benefit. These practices undermine the market's ability to operate efficiently, reduce other market participants' confidence in the markets and distort market outcomes, including prices. These practices are discussed in Chapter 5, Market Manipulation.

Addition Information

This primer is written to be used either as a traditional text – read front to back – or as a reference guide. Consequently, some material is repeated in different sections and references are provided to other parts of the primer where a concept is addressed in greater detail.

Further information about various aspects of energy markets and FERC regulation can be found at www.ferc.gov; then click on Market Oversight.

Google search also provides a quick path to information on specific FERC orders or to more general subjects (e.g., FERC regulation of natural gas pipelines).

2. Wholesale Natural Gas Markets

Overview

Natural gas markets have a significant impact on the economy and on the individuals who rely on the fuel for electric generation, manufacturing, heating, cooking and other purposes. The Department of Energy's Energy Information Administration (EIA) estimates that natural gas supplies 25 percent of the energy used in the United States, or about 24 trillion cubic feet (Tcf) of gas a year.

Under the Natural Gas Act (NGA), the Federal Energy Regulatory Commission (FERC) has jurisdiction over the transportation and sale of natural gas and the companies engaged in those activities.

The natural gas market is an amalgamation of a number of subsidiary markets. There is a physical market, in which natural gas is produced, transported, stored and consumed. There is also a financial market in which physical natural gas is bought and sold as a financial product derived from physical natural gas. Natural gas markets are also regional, with prices for natural gas varying with the demand characteristics of the market, the region's access to different supply basins, pipelines and storage facilities.

Natural Gas

Natural gas is primarily methane, and is colorless and odorless in its natural condition. It is also highly combustible, giving off a great deal of energy and fewer emissions than fuels such as coal and oil. Natural gas occurs in geological formations in different ways: as a gas phase associated with crude oil, dissolved in the crude oil, or as a gas phase not associated with any significant crude oil. Natural gas is rich or wet if it contains significant natural gas liquids (NGL); by contrast, natural gas is lean or dry if it does not contain these liquids. Processors remove water, liquefiable hydrocarbons and other impurities from the natural gas stream to make the natural gas suitable for sale. Natural gas liquids may be processed out and sold separately.

Natural gas is mostly methane, which is made of one carbon atom and four hydrogen atoms (CH_4) and is among the materials known as hydrocarbons.

Other gases such as ethane, butane, propane and pentane may be mixed in with the methane. These may be processed out and sold separately. Gas containing significant amounts of natural gas liquids is known as wet gas. The gas that fl ows through pipelines to natural gas consumers typically has these liquids removed, and is known as dry gas.

While natural gas is typically a gas, it can be cooled to a liquid and transported in trucks or ships. In this form, it is referred to as liquefied natural gas, or LNG.

Natural Gas Markets

As noted, natural gas markets are both physical and financial. This chapter focuses on the physical natural gas markets, but it should be noted that financial markets can have significant impacts on the physical natural gas market.

For this discussion, the natural gas industry has three segments. The first is the supply segment, which includes exploration and development of natural gas resources and reserves, and production, which includes drilling, extraction and gas gathering. The second segment is the midstream sector, in which small-diameter gathering pipeline systems transport the gas from the wellhead to natural gas processing facilities, where impurities and other hydrocarbons are removed from the gas to create pipeline-quality dry natural gas. The third segment is transportation, which includes intrastate and interstate pipeline systems that move natural gas through large-diameter pipelines to storage facilities and a variety of consumers, including power plants, industrial facilities and local distribution companies (LDCs), which deliver the gas to retail consumers.

Each component of the supply chain is critical in serving customers. The quantity of reserves and production can affect market participants' expectations about current and future supply, and thus can affect prices. Similarly, the availability of pipeline and storage capacity determines which supply basins are used and the amount of gas that can be transported from producers to consumers. All of these factors affect the supply chain, but they also affect the supply-demand balance, both nationally and regionally.

Natural gas markets are generally divided into the West, Midwest, Gulf Coast, Northeast and Southeast regions. These regions have differing supply, transportation and demand characteristics, resulting in different prices.

Within these regions are hubs – the interconnection of two or more pipelines – that also become market hubs for buying and selling gas. The key hub used to reflect the U.S. natural gas market as a whole is the Henry Hub, in Louisiana. Prices at other locations are frequently shown as Henry Hub plus or minus some amount.

These regional differences in supply and demand result in different prices for natural gas at various locations. Prices are lowest in areas with low-cost production, ample infrastructure and limited demand – the Opal Hub in Wyoming, for example – and highest where production or transportation is limited and demand is high – Algonquin citygate, in Massachusetts, for example.

Current Trends in Physical Natural Gas Markets

Natural gas markets in the United States are undergoing a period of transition. Within the last decade, various factors have shifted the dynamics of supply and demand. These include, but are not limited to, the following:

1) Development of technology, like hydraulic fracturing and horizontal drilling, enables producers to access unconventional resources such as those in shale formations. This has vastly expanded supply and is increasing the amount of natural gas produced, which has reached levels not seen in more than 35 years. It also has moderated prices across the country. Notably, some of these resources are located close to eastern population centers, providing access to low-cost gas supplies with lower transportation costs.
2) Natural gas has become an investment opportunity as it is a traded commodity. As noted above, there are physical and financial investment markets. There are two distinct markets for physical natural gas: (1) a cash market, which is a daily market where natural gas is bought and sold for immediate delivery; and (2) a forward market, where natural gas is bought and sold under contract for one month or more in the future. The financial gas market is directly linked to the physical natural gas market.
3) Natural gas demand for power generation is rising and is expected to increase significantly in the coming years. Power plant demand for natural gas reflects the environmental benefits of the fuel, the operating flexibility of natural gas-fired generators, and lower natural gas prices. Natural gas-fired power plants emit less air pollution than generators using coal or oil. These plants are also relatively easier to site, can be built in a range of sizes and can increase or decrease output more flexibly than large baseload generators, such as nuclear or coal. This ability to change output

quickly aids electric system operators in matching generation to customer loads, and enables operators to offset rapid changes in output from wind and other intermittent generators.

4) Pipeline expansion has changed the relationships between prices in various regions. New interstate pipelines have enabled regions such as the Northeast and Mid-Atlantic to access new supply sources, expanded the amount of natural gas that can flow from traditional supply sources and enhanced the amount that can flow overall. This has reduced prices and tempered extreme price movements during periods of peak demand.

Natural Gas Demand

Natural gas already is the fuel of choice for many sectors of the U.S. economy. Natural gas demand, however, can fluctuate dramatically, but it generally provides about 25 percent of U.S. energy needs.

Demand Drivers

Over the long term, natural gas use is driven by overall economic and population growth, environmental policy, energy efficiency, technological changes and prices for natural gas and substitute energy sources such as oil, coal and electricity. In the short-term, demand stems from weather, economic activity and changing relationships between coal and natural gas prices.

Weather

Weather is the most significant factor affecting natural gas demand, which has historically been seasonal and weather-driven. Natural gas demand can also swing considerably within a given day, especially during periods of extreme temperatures. Short-term changes in weather, such as heat waves and winter storms, can send demand and prices soaring – or dropping – within the course of a day, sometimes unexpectedly. This unpredictability challenges suppliers and pipelines, especially when pipelines are already full.

Economic Activity and Growth

Economic growth can increase the amount of natural gas used by industry, power plants and commercial entities as consumers want more of their products and services. During a recession, gas use usually declines.

On the other hand, economic growth may raise personal incomes and consumption of electric-powered consumer goods.

Structural changes in the economy can also affect natural gas demand. Declining manufacturing and growing service sectors result in changes in gas use, as does increased global competition. New markets for products and services may require additional natural gas; movement of operations offshore may reduce it.

Daily and weekly economic activity creates cyclical demand patterns. During the work day, demand rises as people get up and go to work or school. Similarly, it declines as they go to sleep. On the weekend, demand tends to vary less over the course of the day.

Prices of Natural Gas and Coal

To use natural gas for heat, a home needs to have an appropriate furnace and piping in place. To burn it in a power plant, the generator needs to make long-term investments in gas-fired generators. These are decisions requiring longterm capital investments, and are cheapest and easiest to make at the time a home or power plant is being built, and are more complicated to change later. Thus, over the long term, demand for natural gas can be affected by the expected costs of alternative energy sources: the cost of a natural gas furnace versus an electric one; the cost of a coal-fired generating plant versus one fueled by natural gas.

In the short-term, the opportunity for fuel switching has been significant in power generation. Electric grid operators have a choice as to which power plant to dispatch to meet increased electric demand. Dispatch is often based on the marginal cost of generation at each available plant in the generation fleet. Plants with lower marginal costs, such as nuclear, typically dispatch before plants with higher marginal costs, such as natural gas.

As natural gas prices drop relative to coal prices, natural gas-fired generation can get dispatched earlier than coal-fired generation, increasing natural gas demand from the power sector.

Demographics and Social Trends

Long-term demand can also be affected by shifting demographics and social trends. Population growth in warmer climates and declines in the older industrial areas of the North have affected natural gas use. So has the trend toward larger houses.

Today, most households have a proliferation of electronic appliances and gadgets. Even as appliances become more energy efficient, people find more appliances to use. Thus, the result of greater efficiency is not necessarily less use of electricity. But one thing for sure is that a greater share of the electricity that is generated is fueled by natural gas.

Environmental Concerns and Energy Efficiency

Natural gas has relatively fewer environmental problems compared with other fossil fuels, and, consequently, it is being increasingly used for power generation. In addition to helping urban areas meet air quality goals, natural gas generation has not experienced as much negative public sentiment as have nuclear and coal-fired generators, making it feasible to site gas-fired generators closer to load centers. Growth in wind and other intermittent generation technologies benefit when coupled with natural gas generation, which is able to ramp up and down quickly to complement variable output by wind.

The natural gas emissions profile has also encouraged some urban mass transit bus systems, West Coast port operations and other vehicle fleets to shift to natural gas from gasoline or diesel fuel.

Customer Sectors and Demand

In 2007, natural gas used for electric generation overtook gas-for-industrial load to become the largest customer class for natural gas. In 2010, according to the EIA, power generation used 7.4 Tcf of the natural gas delivered to consumers; industrial, 6.6 Tcf; residential, 4.8 Tcf; and commercial, 3.2 Tcf.

Demand is seasonal. It rises and falls during a year as the seasons bring changing temperatures. Demand usually peaks in winter to meet heating load, with a second, smaller peak in summer to meet power generator load. The low months for gas consumption are during spring and fall, known as shoulder months. Natural gas use also varies with the time of day. The difference between peak winter and off-peak demand varies regionally, due to differences in winter temperatures and customer composition.

Each customer sector contributes differently to overall demand, both in terms of the amount that demand varies over a cycle and whether its peak demand coincides with the overall system peak. Residential demand, for example, can be highly variable in colder climates, and its peak coincides with the overall system peak. Power generation's peak does not coincide with the winter gas-demand peak, but in fact its growing use of natural gas to produce electricity for air conditioning has created robust summer demand, which competes with gas supply that traditionally would flow into underground storage for later use in the winter. Industrial demand stays relatively constant year-round.

In the short term, residential and commercial natural gas use tends to be inelastic – consumers use what they need regardless of the price. Power plant demand, on the other hand, is more price-responsive as natural gas competes with other fuels, especially coal, in the production of electricity. Price inelasticity implies that a potential for price spikes exists during periods of supply constraint.

Consequently, the mix of customers in a region can affect system operations and costs. Pipelines and other equipment need to be sized to account for peak demand. Demand that stays fairly constant presents fewer operational challenges and usually enjoys lower prices. Highly variable demand will result in pipelines and equipment being used at less than full capacity for much of the year, and prices for service may be more expensive, both because the pipelines may become constrained during peak times and because the capacity is not fully used throughout the year.

Power Generation

Generation demand can soar at any time; gas-fired generators can change their output quickly, and are frequently called on to change their output due to changes in demand or when problems occur elsewhere in the power grid. United States natural gas demand for power generation increased from 5.5 Tcf in 2004 to 7.4 Tcf in 2010, according to the EIA. Generating plants tend to consume more natural gas in the summer to meet air conditioning loads, but power demand can also climb in the winter to provide electric heating and lighting.

Generation demand can also be influenced by the relative prices for natural gas and other fuels, especially coal. Since late 2008, natural gas-fired generators generally have been dispatched before some coal plants because of the decrease in natural gas prices.

Industrial

Natural gas as a fuel is used to produce items such as steel, glass, paper, clothing and brick. It also is an essential raw material for paints, fertilizer, plastics, antifreeze, dyes, photographic film, medicines and explosives. Industrial load tends to show the least seasonal variation of natural gas use, but the industry is sensitive to economic pressures. Gas use has been declining slowly due to efficiency improvements and economic factors, such as the movement of manufacturing overseas. Still, EIA data show that industrial customers remain the second-largest consuming segment, using 6.6 Tcf in 2010, down from 7.2 Tcf in 2004. This decline stemmed from both fewer customers and lower average use.

Residential

Despite population growth, natural gas used in the residential sector for home furnaces, water heaters, clothes dryers and stoves has remained fairly flat over the past decade as appliances and homes have become more energy efficient. Residential customers consumed 4.8 Tcf in 2010, compared to 4.9 in 2004. Much of the year-to-year demand variation in this sector can be attributed to the weather during a particular year. A year with a long, cold winter and hot summer will see higher gas demand than a year with a mild winter and summer. Demographics may also support this trend, as customers have moved from northern industrial centers to warmer parts of the country. Slightly more than half of the homes in the United States use natural gas as their main heating fuel. Residential customers typically show the greatest seasonal variability, especially in cold-winter regions where demand soars during winter months as consumers turn on their furnaces.

Commercial

Commercial demand has remained fairly constant at 3.2 Tcf in 2010 as well as in 2004, and like the residential sector will see some year-to-year variation based on weather. Commercial consumers include hotels, restaurants, wholesale and retail stores and government agencies, which use natural gas primarily for heat. Consequently, its demand cycles over the seasons, weeks and days.

Quick Facts: Resources and Proved Reserves

- Resources - Total gas estimated to exist in a particular area. The Potential Gas Committee's latest biennial estimate showed 2,170 Tcf of future gas supply in the United States as of Dec. 31, 2010; that figure included 273 Tcf of proved dry gas reserves and 1,898 Tcf of total potential resources.
- Proved reserves - Estimated amount of natural gas that, based on analysis of geologic and engineering data gathered though drilling and testing, can be reasonably projected to be recoverable under existing economic and operating conditions. According to the EIA, as of Dec. 31, 2009, dry natural gas proved reserves totaled 273 Tcf. Proved reserves have been growing every year since 1999.

Natural Gas Supply

Natural Gas Resources, Reserves and Production

The amount of natural gas in the ground is estimated by a variety of techniques, and it takes into account the technology available to extract the gas. According to the EIA, estimating the technically recoverable oil and natural gas resource base in the United States is an evolving process. Analysts use different methods and systems to make natural gas estimates. As detailed in the table below, natural gas supplies are characterized as resources, proved reserves and production.

Resources is the largest category, which describes the total potential of natural gas supply. Proved reserves consider the feasibility and economics of extracting the natural gas. Lastly, production describes the amount of natural gas removed from the ground.

Natural gas is located underneath the surface of the earth. Natural gas is characterized by the type of basin or rock formation in which it lies. Conventional natural gas is found in porous rock formations, and in the United States is the traditional source of natural gas.

Unconventional natural gas, on the other hand, is found in shale, coal seams and tight, low-permeability rock formations. In 2007, the National Petroleum Council (NPC) defined unconventional gas as "natural gas that cannot be produced at economic flow rates or in economic volumes of natural gas unless the well is stimulated by a large hydraulic fracture treatment, a horizontal wellbore or by using multilateral wellbores or some other technique to expose more of the reservoir to the wellbore."

In the past few years, improvements in drilling technology have enabled producers to access unconventional supplies, notably shale, yielding significant increases in production and raising the estimate of proved reserves. Estimates of resources in 2011 amounted to approximately 2,170 Tcf (which included reserves).

This domestic growth in resources and reserves has translated into greater natural gas production, which has grown 22 percent since 2005, to more than 60 billion cubic feet per day (Bcfd) in 2011. Most of the growth came from shale gas, which now accounts for 32 percent of natural gas resources.

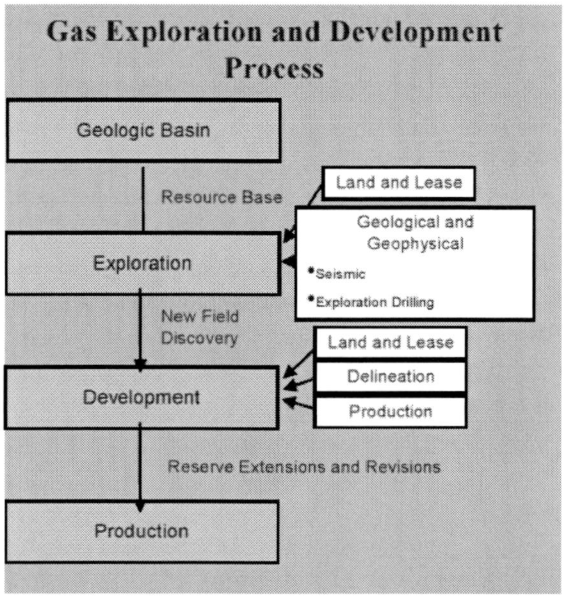

Worldwide, the United States accounts for one-third of global natural gas resources. Most of the natural gas resources are in the Middle East – Iran, Qatar and Saudi Arabia – followed by the United States and Russia.

Rig Count and Rig Productivity

A measure of exploration, the rig count measures the number of rotary drilling rigs actually drilling for oil and gas. These measures are done by several companies active in drilling operations. Rig counts are often used as a rough predictor of future production. The oil and gas rig count peaked at 4,530 on Dec. 28, 1981. More recently, in September 2008, the rig counted reached 2,500, before plunging to 1,200 in April 2009, according to Baker Hughes Inc. Rig counts recently have hovered around 1,900.

Within the total rig count, the use of horizontal drilling rigs, capable of accessing natural gas and oil in shale formations, grew prior to the recession that began in 2008. Traditional vertical rigs remain well below their prerecession highs.

The adoption of horizontal drilling has significantly increased production per rig, making comparison of rig counts over time problematic because horizontal rigs are considerably more productive than vertical rigs.

Conventional and Unconventional Natural Gas

Natural gas is a fossil fuel. Natural gas historically has been found in underground reservoirs made when organic material was buried and pressurized. The remains of that organic material were trapped in the surrounding rock as oil or natural gas. Natural gas and oil are often found together. The depth of the organic materials and the temperatures at which they are buried often determine whether the organic matter turns into oil or natural gas. Generally, oil is found at depths of 3,000 to 9,000 feet; organic materials at greater depths and higher temperatures result in natural gas.

Natural gas basins are frequently referred to as conventional or unconventional basins or plays. These basins differ in the geology of the basin and the depth at which gas can be found. The schematic at left illustrates differing geologic formations in which natural gas can be found.

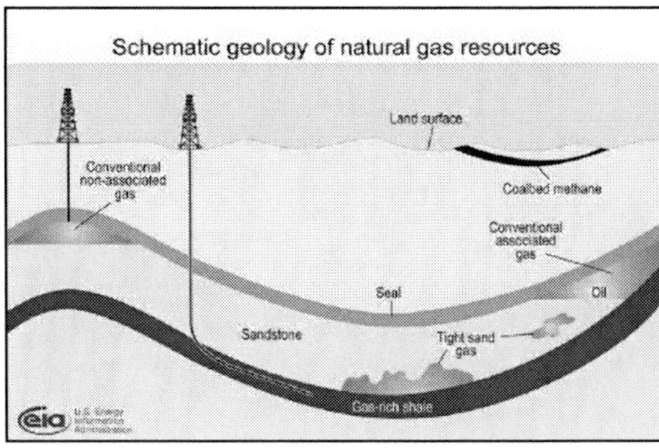

Conventional Natural Gas

Natural gas historically has been produced from what is traditionally known as conventional natural gas resources, which provided most of the country's supply needs for more than a century. Conventional gas is found in geological basins or reservoirs made of porous and permeable rocks, holding significant amounts of natural gas in the spaces in the rocks. Conventional resources have been found both on land and offshore (see map of conventional fields, on next page), with the major fields in an arc from the Rocky Mountains to the Gulf of Mexico to Appalachia. The largest conventional fields reside in Texas, Wyoming, Oklahoma, New Mexico and the federal offshore area of the Gulf of Mexico. In 2000, offshore natural gas production represented 24 percent percent of total U.S. production; by 2010 that amount had fallen to 10 percent.

Gas Production in Conventional Fields, Lower 48 States

Source: Energy Information Administration based on data from HPDI, In Geological Survey, USGS. Updated April 8, 2009.

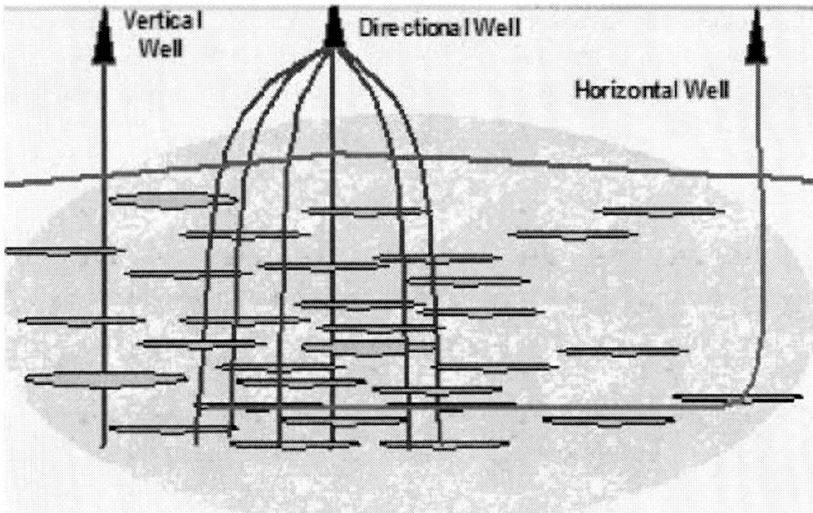

Source: Energy Information Administration.

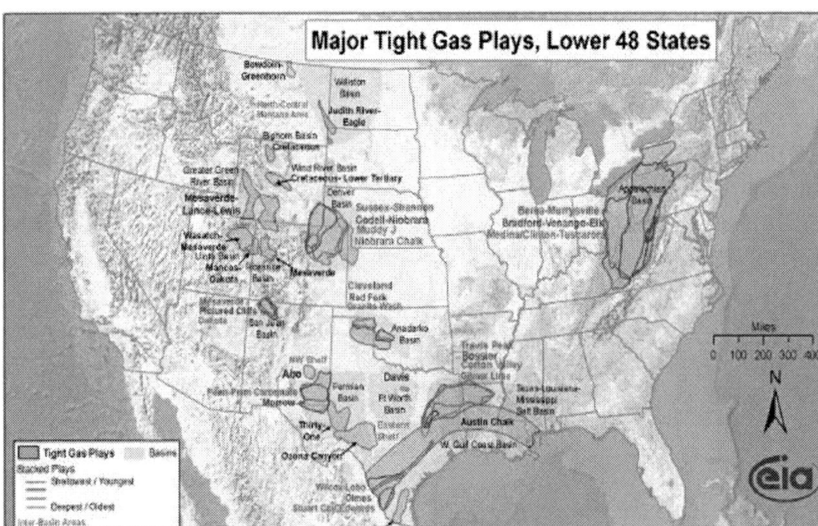

Source: Energy Information Administration based on data from various published studies, Updated June 6, 2010.

Federal offshore natural gas wells are drilled into the ocean floor off the coast of the United States in waters that are jurisdictional to the federal government. Most states have jurisdiction over natural resources within three nautical miles of their coastlines; Florida and Texas claim nine nautical miles of jurisdiction.

Roughly 4,000 oil and gas platforms are producing in federal waters at water depths approaching 7,500 feet (at total well depths of 25,000-30,000 feet) and at distances as far as 200 miles from shore, EIA reports. Most of these offshore wells are in the Gulf of Mexico.

Offshore production has been going on for decades. As the easy pickings – close-in, shallow-water wells – became less economic to produce, companies looked to reserves at greater water depth. Technological improvements contributed to continuing production from deep offshore wells.

Unconventional Natural Gas

In recent years, innovations in exploration and drilling technology have led to rapid growth in the production of unconventional natural gas. This term refers to three major types of formations where gas is not found in distinct basins, but is trapped in shale, tight sands or coal seam formations over large areas.

Coalbed Methane Fields, Lower 48 States

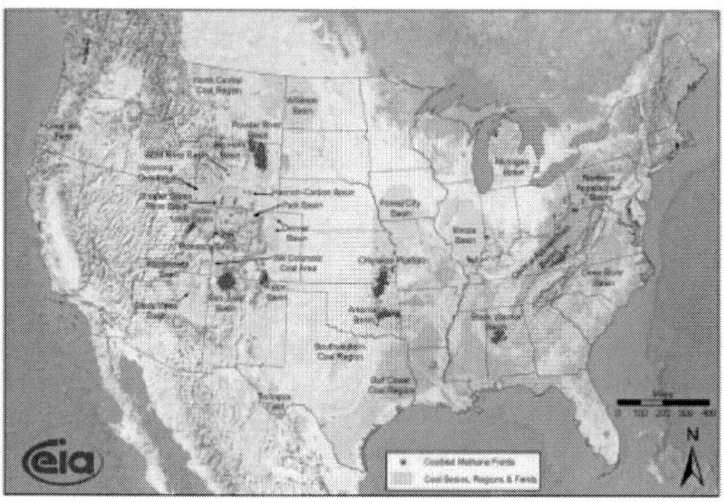

Source: Energy Information Administration based on data from USGS and various published studies, Updated April 8, 2009.

The presence of natural gas in these unconventional plays has been common knowledge for decades, but it was not until the early 1990s, when after years of experimenting in the Barnett Shale in Texas, George Mitchell and Mitchell Energy Co. developed a new drilling technique that made production in these types of formations economically feasible. The new technology included horizontal and directional wells, which allow a producer to penetrate diverse targets and increase the productivity of a well. As shown below, directional wells allow the producer to tap these resources through multiple bores. The horizontal wells have a vertical bore, but then move horizontally through the rock to access more supply. These new drilling technologies greatly improved the likelihood of a successful well and the productivity of that well.

As of 2012, production from unconventional reserves supplied more than half of U.S. gas needs.

Tight sands gas is natural gas contained in sandstone, silt-stone and carbonate reservoirs of such low permeability that it will not naturally flow when a well is drilled. To extract tight sands gas, the rock has to be fractured to stimulate production. There are about 20 tight sands basins in the United States (see map); as of 2009, annual production was about 6 Tcf, or about one-third of U.S. domestic production.

Coalbed methane (CBM) is natural gas trapped in coal seams. Fractures, or cleats, that permeate coalbeds are usually filled with water; the deeper the coalbed, the less water is present. To release the gas from the coal, pressure in the fractures is created by removing water from the coalbed.

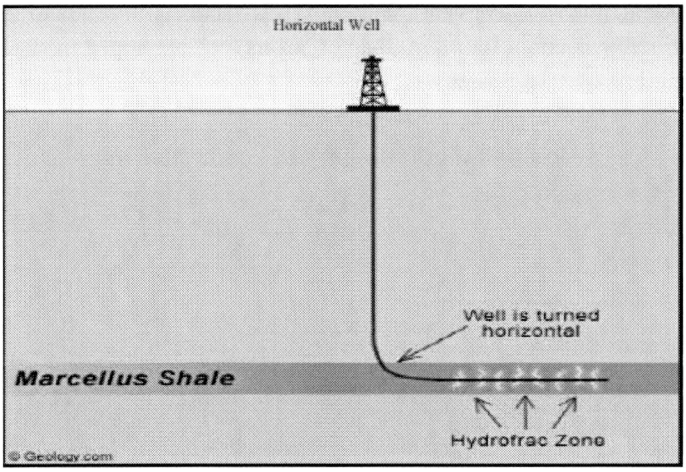

The in-place coalbed methane resource of the United States is estimated to be more than 700 Tcf, but less than 100 Tcf of that may be economically recoverable, according to the U.S. Geological Survey. Most CBM production in the United States is concentrated in the Rocky Mountain area, although there is significant activity in the Midcontinent and the Appalachian area.

Shale gas is found in fine-grained sedimentary rock with low permeability and porosity, including mudstone, clay stone and what is commonly known as shale. These rock conditions require a special technique known as hydraulic fracturing (fracking) to release the natural gas. This technique involves fracturing the rock in the horizontal shaft using a series of radial explosions and water pressure (see graphic).

In the past decade the processes for finding geological formations rich in shale gas, or shale plays, have improved to the point that new wells almost always result in natural gas production. Improved exploration techniques coupled with improved drilling and production methods have lowered the cost of finding and producing shale gas, and have resulted in a significant increase in production. In 2011, shale gas accounted for about 25 percent of total gas production, with expectations of significant increases in the future.

The five major shale plays in the United States include Barnett, Fayetteville, Woodford, Haynesville and Marcellus (see map on next page). Together, they hold an estimated 3,420 Tcf of shale gas resources, according to Bentek Energy. Other shale formations, such as Eagle Ford, are seeing heavy exploration activity and are expected to become major contributors of natural gas supply in the near future. The shale plays are widely distributed through the country, which has the added advantage of putting production closer to demand centers, thus reducing transportation bottlenecks and costs.

Many shale reservoirs contain natural gas liquids (NGL), which can be sold separately, and which augment the economics of producing natural gas. Natural gas is rich or wet if it contains significant liquids; lean or dry if is does not.

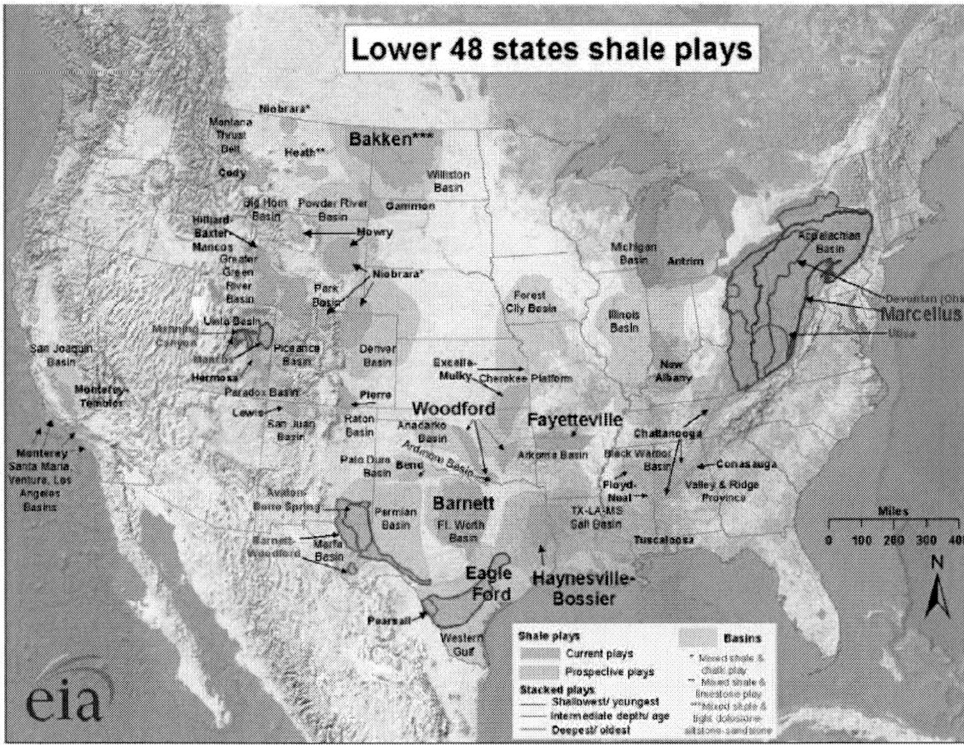

Source: Energy Information Administration based on data from various published studies, Updated May 9, 2011.

The Shale Revolution

The estimated resources, proven reserves and production of shale gas have risen rapidly since 2005, and shale is transforming gas production in the United States. In 2009, according to EIA, shale gas made up 13 percent of gross production of natural gas, and is expected to become the dominant source of domestically produced gas. By comparison, coalbed methane accounted for 8 percent of production, while 22 percent of the natural gas came from oil wells and 57 percent was produced from natural gas wells.

New shale plays have increased dry shale gas production from 1 Tcf in 2006 to 4.8 Tcf, or 2 percent of total United States natural gas production, in 2010. Wet shale gas reserves account for about 21 percent of the overall United States natural gas reserves. According to the EIA, shale gas will account for about 48 percent of United States natural gas production in 2035. Shale gas can be produced at much lower costs, overall, than gas from conventional fields. Breakeven prices – the price that equals production cost – are below recent natural gas prices, making production of shale gas potentially profitable. The natural gas liquids add to the value of production.

Shale Wet Gas Breakeven Prices

Natural gas liquids are priced more like oil than natural gas, making them lucrative. A typical barrel of NGL might contain 40-45 percent ethane, 25-30 percent propane, 5-10 percent butane and 10-15 percent natural gasoline.

As a result, the higher value of the liquids coming up the well offset the lower market prices for the dry gas portion of the wellstream.

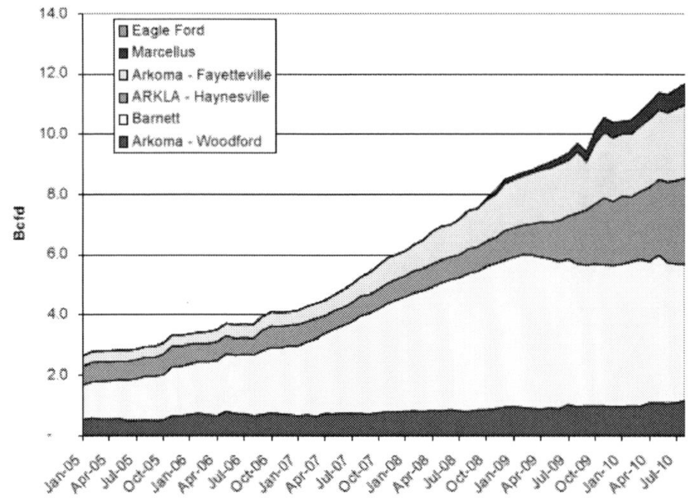

Source: Derived from Bentek Energy data.

Shale Gas Production by Region.

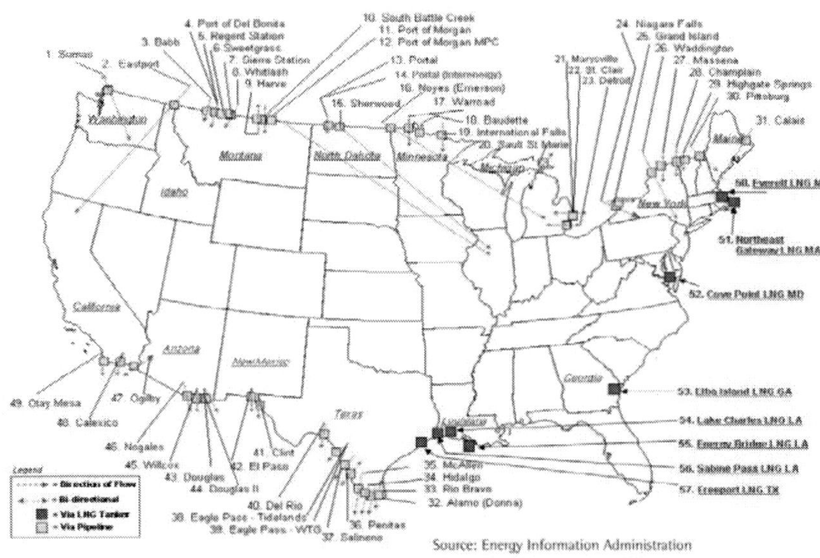

Natural Gas Import Entry Points.

In addition to having lower finding and production costs, shale gas production is more flexible than traditional production. Unless the well is plugged, a vertical well is always producing. Shale gas production, however, allows the producer to essentially turn off the well without endangering the well by slowing the pace of hydraulic fracturing along the horizontal bore hole. Therefore, producers are able to tailor shale production to market conditions. The Marcellus Shale formation in Appalachia is of particular note because of its location, size and resource potential, according to the Potentail Gas Committiee at the Colorado School of Mines. Marcellus Shale has estimated gas resources reaching 549 Tcf, and it extends from West Virginia to New York, near the high population centers of the Northeast and Mid-Atlantic. Although Marcellus Shale has been producing significant amounts of gas only since 2008, production has been prolific with high initial well pressures and high production rates.

Growing gas production in Marcellus has already made an impact on U.S. gas transportation. As more gas has flowed out of Marcellus, less gas has been needed from the Rockies or the Gulf to serve the eastern United States. This new production has contributed to a reduction in natural gas prices and the long-standing price differentials between the Northeast and other parts of the United States. It has also caused imports from Canada to decrease.

Environmental concerns present the greatest potential challenge to continued shale development. One issue involves the amount of water used for hydraulic fracturing and the disposal of the effluent used – chemicals and sand are combined with water to create a fracturing solution, which is then pumped into deep formations. Some companies recycle the returned water, which allows them to reuse such water.

Concerns have also been raised regarding the potential risks and health hazards associated with wastewater (especially when stored at ground level in holding ponds) seeping into drinking water.

FERC Jurisdiction

Section 1(b) of the NGA exempts production and gathering facilities from FERC jurisdiction. Moreover, the Wellhead Decontrol Act of 1989, Pub. L. No. 101-60 (1989); 15 U.S.C. § 3431(b)(1)(A), completely removed federal controls on new natural gas, except sales for resale of domestic gas by interstate pipelines, LDCs or their affiliates.

In Order No. 636, FERC required interstate pipelines to separate, or unbundle, their sales of gas from their transportation service, and to provide comparable transportation service to all shippers whether they purchase gas from the pipeline or another gas seller.

Imports and Exports

Natural gas imports play an important role in regional U.S. markets, accounting for about 3,800 Bcf, or 11 percent, of the natural gas used in the United States in 2011. The natural gas pipeline systems of the United States and Canada are integrated, and about 90 percent of imports came from Canada, according to the EIA, while 10 percent was imported as liquefied natural gas (LNG).

Imported natural gas flows into the United States via pipelines at numerous points along the U.S. border with Canada. Imports from Canada have been of strategic importance in the Northeast and the West, which were traditionally far from the major domestic production centers. However, Canadian exports to the United States have been declining as U.S. shale production has increased. Net U.S. gas imports have declined from a recent high of 3,785 Bcf in 2007 to 1,948 Bcf in 2011. EIA estimates that imports will continue to decrease as shale-gas production increases.

The United States also exports natural gas to Canada and Mexico, and it still occasionally exports LNG to Japan.

Liquefied Natural Gas

Liquefied natural gas (LNG) is natural gas cooled to minus 260 degrees Fahrenheit to liquefy it, which reduces its volume by 600 times. LNG may be transported in ships and trucks to locations not connected by a pipeline network.

FERC Jurisdiction

The FERC has exclusive authority under the NGA to authorize the siting of facilities for imports or exports of LNG. This authorization, however, is conditioned on the applicant's satisfaction of other statutory requirements for various aspects of the project. In addition, the Department of Energy has authority over permits to import and export.

The LNG Supply Chain

Natural gas is sent to liquefaction facilities for conversion to LNG. These facilities are major industrial complexes, typically costing $2 billion, with some costing as much as $20 billion.

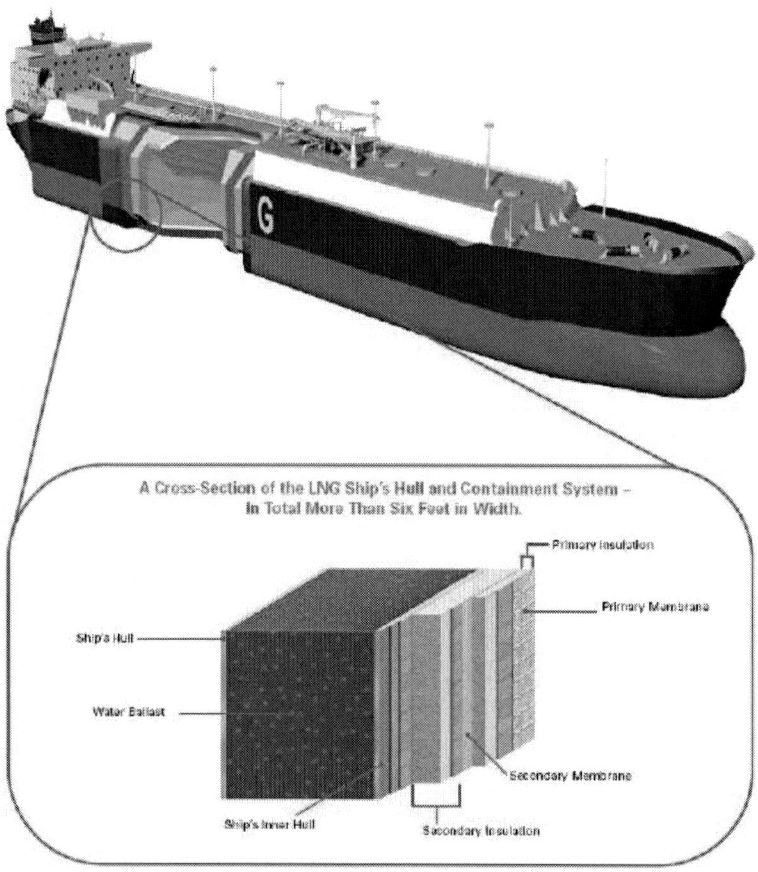

Once liquefied, the LNG is typically transported by specialized ships with cryogenic, or insulated, tanks.

Once LNG reaches an import (regasification) terminal, it is unloaded and stored as a liquid until ready for sendout. To send out gas, the regasification terminal warms the LNG to return it to a gaseous state and then sends it into the pipeline transportation network for delivery to consumers. Currently, more than 80 Bcfd of regasification capacity exists globally, more than double the amount of liquefaction capacity. Excess regasification capacity provides greater flexibility to LNG suppliers, enabling them to land cargoes in the highest-priced markets. This flexibility has fostered a growing spot market for LNG.

The cost of the LNG process is \$2-\$4 per million British thermal units (MMBtu), depending on the costs of natural gas production and liquefaction and the distance over which the LNG is shipped. Liquefaction and shipping form the largest portion of the costs. Regasification contributes the least cost of any component in the LNG supply chain. The cost of a regasification facility varies considerably; however, the majority of these costs arise from the development of the port facilities and the storage tanks. A 700-MMcfd terminal may cost in the range of \$500 million to \$800 million.

The various components of the LNG process are broken out below.

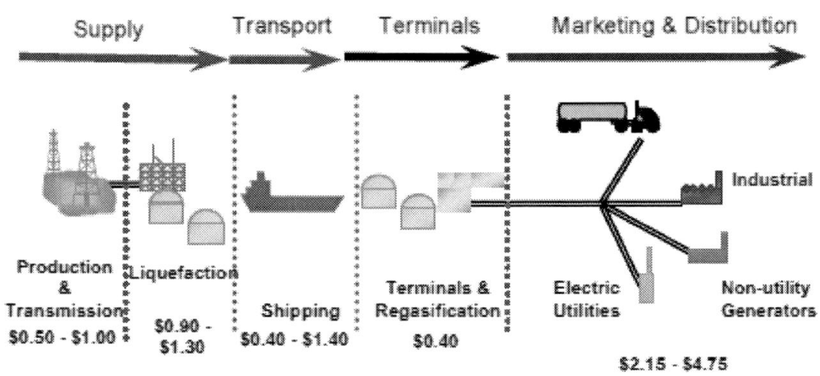

Source: Drewry Shipping.

LNG in the United States

The United States is second to Japan in LNG regasification capacity. As of 2011, there were 11 LNG receiving or regasification terminals in the continental United States, with approximately 13 Bcfd of import capacity and 79 Bcf of storage capacity. All of these facilities are on the Gulf or East coasts, or just offshore. In addition, the United States imports regasified LNG into New England from the Canaport LNG terminal in New Brunswick, Canada, and into Southern California from the Costa Azul LNG terminal in Mexico's Baja California.

Between 2003 and 2008, the United States met 1-3 percent of its natural gas demand through LNG imports, according to the EIA. LNG imports peaked at about 100 Bcf/ month in summer 2007. Growth in relatively low-cost U.S. shale gas production has trimmed U.S LNG imports, affecting Gulf Coast terminals the most. Today, most LNG enters the United States under long-term contracts (about half of the total) coming through the Everett (Boston) and Elba Island (Georgia) LNG terminals. The remainder of the LNG enters the United States under short-term contracts or as spot cargoes. LNG prices in the United States generally link to the prevailing price at the closest trading point to the import terminal. During 2011-12, the growth in shale gas production led to proposals to export significant volumes of domestically produced LNG. Since 1969, small quantities of LNG have been shipped from Alaska to Pacific Rim countries.

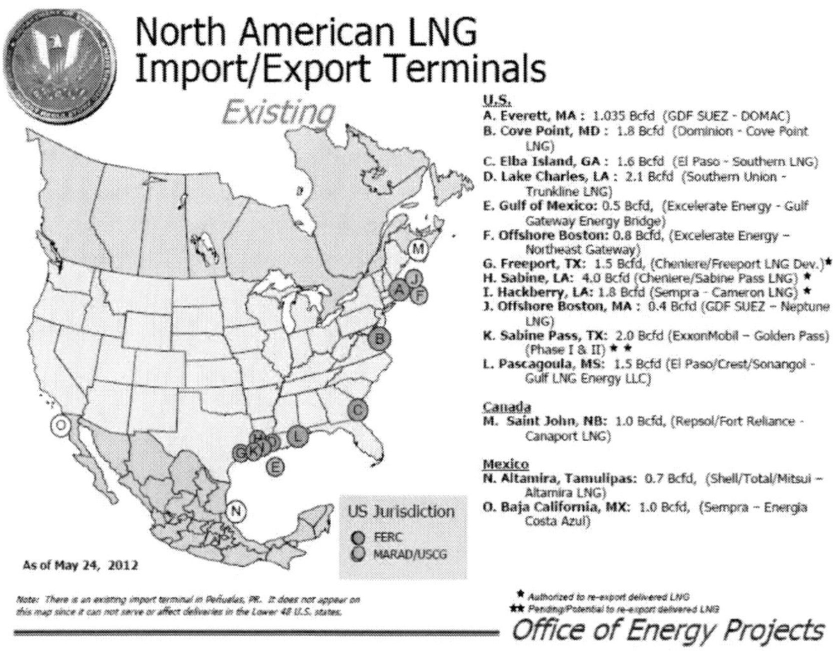

Natural Gas Processing and Transportation

Most domestic natural gas production in the United States occurs in regions well away from major population centers.

To get gas from wellhead to consumers requires a vast network of processing facilities and 2.4 million miles of pipelines. In 2010, this network delivered more than 22 Tcf of natural gas to millions of customers. The U.S. natural gas system can get natural gas to and from almost any location in the Lower 48 states.

Efficient markets require that this network be robust and allow consumers access to gas from more than one production center. Supply diversity tends to improve reliability and moderate prices, while constraints increase prices.

Processing

The midstream segment of the natural gas industry between the wellhead and pipelines is shown in the graphic below. This segment involves gathering the gas from the wellhead, processing the gas to remove liquids and impurities and moving the processed (dry) natural gas to pipelines and the extracted liquids to a fractionator that separates the liquids into individual components. The liquids are used by the petrochemical industry, refineries and other industrial consumers. There were about 500 gas processing plants operating in the United States in 2009.

The composition of raw, or wellhead, natural gas differs by region. Consequently, processing will differ depending on the quality of the natural gas. Natural gas may be dissolved in oil underground but separated out from the oil as it comes to the surface due to reduced pressure. In these instances, the oil and gas are sent to separate processing facilities. Where it does not separate naturally, processing is required.

Midstream Overview

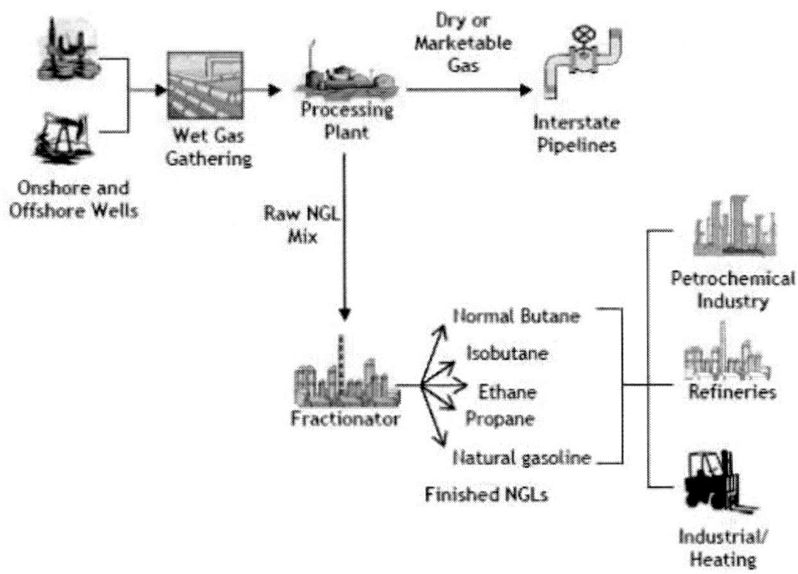

Untreated natural gas usually contains liquids and is generally known as wet gas. Wet gas may contain water, water vapor, hydrogen, propane, butane, natural gasoline and other components, many of them valuable in other consumer and industrial applications. Gas varies widely in the amounts of these liquids it contains.

Once a well is completed and gas is flowing, the gas moves into gathering pipelines, which typically are small-diameter lines that move the gas from the wellhead to either a processing plant or a larger pipeline.

Processing is required when the natural gas and oil do not separate naturally. At the processing plant, wet natural gas is dehydrated, and additional products and contaminants (such as sulfur and carbon dioxide) are extracted. The hydrocarbon liquids extracted are known as natural gas liquids (NGL). Many of these are high-value products used in petrochemical applications. Once processing extracts the NGL, the stream is separated into individual components by fractionation, which uses the different boiling points of the various hydrocarbons to separate them. Once processing is completed, the gas is of pipeline quality and is ready to be moved by intrastate and interstate pipelines.

FERC Jurisdiction

The Natural Gas Act (NGA), 15 U.S.C. § 717 et. seq., gives the FERC comprehensive regulatory authority over companies that engage in either the sale of natural gas for resale or its interstate transportation. The Commission regulates market entry through Section 7 of the NGA, 15 U.S.C. § 717f, by issuing certificates of public convenience and necessity, subject to such conditions as the Commission deems appropriate, authorizing natural gas companies to transport or sell natural gas. To this end, the FERC reviews applications for the construction and operation of interstate natural gas pipelines. In its application review, the FERC ensures that the applicant has certified that it will comply with Department of Transportation safety standards. The FERC has no jurisdiction over pipeline safety or security, but actively works with other agencies with safety and security responsibilities. The Commission regulates market exit through its authority to abandon certificated service, 15 U.S.C. § 717f(b).

Natural Gas Transportation

Interstate pipelines account for 71 percent of the natural gas pipeline miles in the United States and carry natural gas across state boundaries. Intrastate pipelines account for the remaining 29 percent, and have similar operating and market characteristics.

The interstate network moves dry natural gas from producing areas to local distribution companies (LDCs), large industrial customers, electric power plants and natural gas storage facilities. The pipelines, which range in diameter from 16 inches to as large as 48 inches, move gas between major hubs to lateral lines. Laterals, which range in diameter from 6 inches to 16 inches, distribute gas to retail customers.

The large pipelines are known as mainline transmission pipelines. The line pipe used for major pipelines typically consists of strong carbon steel sufficient to meet standards set by the American Petroleum Institute. Line pipe is coated to reduce corrosion. Smaller distribution lines, which operate under much lower pressures, may be made of plastic materials, which provide flexibility and ease of replacement.

Nearly one-fifth of all natural gas transmission pipelines, by mileage, are located in Texas. More than half are located in nine states: Texas, Louisiana, Kansas, Oklahoma, California, Illinois, Michigan, Mississippi and Pennsylvania.

Compressor stations, located every 50-100 miles along the pipe, add to or maintain the pressure of the natural gas, propelling it down the pipeline. Natural gas travels through pipelines at high pressures, from 200 pounds per square inch (psi) to 1,500 psi.

The gas is compressed by turbines, motors or engines. Turbines and reciprocating natural gas engines use some of the gas from the line to fuel their operations; electric motors rely on electricity.

Metering stations are placed along the pipelines to measure the flow of natural gas as it moves through its system.

Movement of natural gas along a pipeline is controlled in part by a series of valves, which can be opened to allow the gas to move freely or closed to stop gas flow along a section of pipe. Large valves may be placed every 5 to 20 miles along the pipeline.

Pipeline operators use supervisory control and data acquisition (SCADA) systems, to track the natural gas as it travels through their systems. SCADA is a centralized communication system that collects, assimilates and manages the meter and compressor station data. SCADA also conveys this information to the centralized control station, allowing pipeline engineers to know what is happening on the system at all times.

As the product moves closer to the consumption areas, it may be stored in underground facilities. Plentiful storage capacity adds flexibility to the pipeline and distribution systems and helps moderate prices by providing an outlet for excess gas during periods of low demand, and readily accessible supply in periods of high demand. Some natural gas can also be stored in the pipelines as linepack, in which more molecules of gas are held in a segment of pipeline under greater-than-normal pressure.

Natural Gas Infrastructure

The United States natural gas market is accommodated by extensive infrastructure:

- Roughly 306,000 miles of wide-diameter, high pressure inter- and intrastate pipelines make up the mainline pipeline transportation network, run by more than 210 companies.
- More than 1,400 compressor stations maintain pressure on the natural gas pipeline network.
- More than 5,000 receipt points, 11,000 delivery points and 1,400 interconnection points implement the flow of gas across the United States.
- Nearly three dozen hubs or market centers provide addition interconnections.
- 400 underground natural gas storage facilities increase the flexibility of the system.
- 49 locations enable natural gas to be imported or exported via pipelines.
- There are 11 LNG import facilities and100 LNG peaking facilities (stored gas held for peak demand periods).
- More than 1,300 local distribution companies deliver natural gas to retail customers.

Source: EIA.

Hubs

A key part of the pipeline distribution network is the natural gas hub. Typically, a hub is a specific point where pipeline interconnections allow the transfer of gas from one pipeline to another.

There are dozens of natural gas hubs in the country, with about 20 considered major hubs. Henry Hub is the dominant benchmark point in the physical natural gas market because of its strategic location in the Gulf Coast's producing area and the number of pipeline connections to the East Coast and Midwest consumption centers. It sits in southcentral Louisiana, in the town of Erath, where more than a dozen major natural gas pipelines converge and exchange gas. The Henry Hub has 12 delivery points and 4 major receipt points.

Gas as a physical product can be bought and sold at Henry Hub or other hubs around the country in daily and monthly markets. In addition, the New York Mercantile Exchange (Nymex) established a futures contract centered at the Henry Hub in 1990 that gained widespread acceptance and is generally used as the reference price for natural gas in the United States.

Distribution lines typically take natural gas from the large transportation pipelines and deliver the gas to retail customers. While some large consumers – industrial and electric generation, for example – may take service directly off a transmission pipeline, most receive their gas through their local gas utility, or local distribution company (LDC). These companies typically purchase natural gas and ship it on behalf of their customers, taking possession of the gas from the pipelines at local citygates and delivering it to customers at their meters. This distribution involves a network of smaller pipelines – more than two million miles, according to the U.S. Department of Transportation.

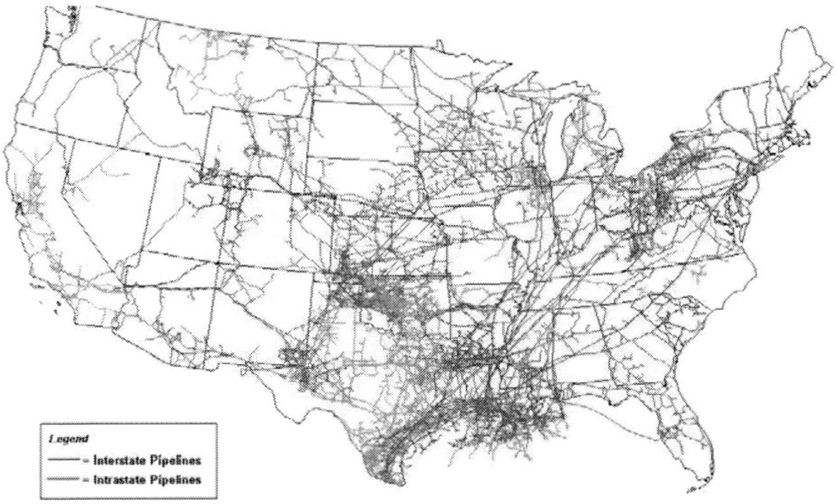

Source: Energy Information Administration, Office of Oil and Gas, Natural Gas Division, Gas Transportation Information System.

Natural Gas Transportation System.

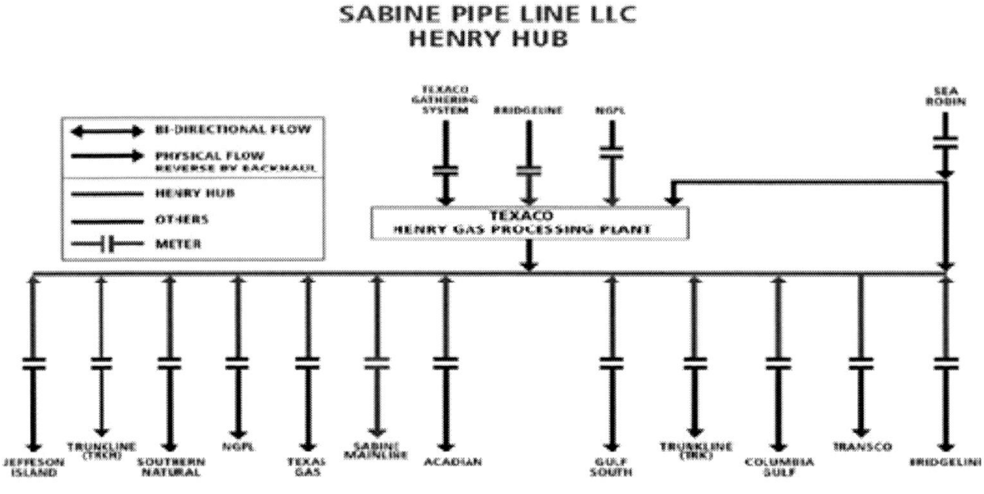

FERC Jurisdiction

The NGA, 15 U.S.C. § 717, et seq., requires that natural gas companies charge just and reasonable rates for the transportation and sale of natural gas. To promote compliance with this mandate, the NGA requires gas pipelines to file rate schedules with the FERC and to notify the FERC of any subsequent changes in rates and charges. On submission of a tariff revision, the FERC may hold a hearing to determine whether the pipeline has met its burden to show that the amended rates and charges are just and reasonable.

Under Sections 4 and 5 of the NGA, 15 U.S.C. §§ 717c and 717d, the Commission regulates the rates and other terms of jurisdictional sales and transportation, ensuring that rates and charges for such service, as well as all rules, regulations, practices, and contracts affecting those rates and charges, are just and reasonable and not the product of undue discrimination. 15 U.S.C. §§ 717c(a) and (b).

Pipeline Services

Customers or shippers have a choice between two general types of service on interstate pipelines. The first is firm transportation capacity, or primary market service, in which an agreement is executed directly between the pipeline and a customer for a year or more, relying on primary receipt and delivery points.

Firm transportation service is not bumped for other classes of service, and it receives the same priority as any other class of firm service.

The second type of transportation service a shipper can contract for is interruptible transportation service. Interruptible transportation service is offered to customers under schedules or contracts on an as-available basis.

This service can be interrupted on a short notice for a specified number of days or hours during times of peak demand or in the event of system emergencies. In exchange for interruptible service, customers pay lower prices.

A secondary market for firm transportation rights enables shippers to sell their pipeline capacity to a third party through the FERC's capacity release program. Services offered in the primary market can be offered in the secondary market by the holder of the primary service. Released capacity offers market participants the opportunity to buy and sell from each other as well as from the pipeline.

This can be broken down into segments: holders of primary capacity can release segments rather than their full holdings, provided segmentation is operationally feasible on the interstate pipeline's system.

Interstate pipelines also provide "no-notice service" under which firm shippers may receive delivery up to their firm entitlements on a daily basis without penalty. If a shipper has firm storage and transportation service, that shipper can schedule in the day-ahead market and yet have the ability and the right to physically take a different quantity than what was scheduled without incurring imbalance penalties.

No-notice service is particularly valuable during periods of high demand when transportation capacity may be completely used. This service is especially helpful to LDCs that must serve their load without knowing their exact load level each day. No-notice service is generally priced at a premium to firm transportation service. Shippers may temporarily release this service to other parties, using FERC-approved capacity release guidelines.

Interstate Transportation Rates

Pipeline transportation rates can be priced on zones or miles, or be a fixed postage stamp rate. In zonal pricing, the price of transportation varies by the location of the receipt and delivery points, across a series of zones.

Under postage stamp rates, shippers pay the same rate for transportation regardless of how far the gas is moved, similar to the way a postage stamp costs the same amount regardless of whether a letter is sent to New York or California. Pipelines using postage stamp rates include Northwest Pipeline, Colorado Interstate Gas and Columbia Gas Transmission.

With mileage-based rates, shippers pay based on the distance between where the gas enters the pipeline and where it is taken out of the pipeline. The rate is designed to reflect the distance involved in transporting the gas. Gas Transmission Northwest (GTN) uses mileage-based rates.

Other pipelines use hybrid or mixed-rate systems. Northern Natural Gas, for example, uses a combination zonal rate for upstream receipts and a postage stamp rate for market area deliveries.

Scheduling

Pipelines have rigorous schedules that shippers must follow. Typically, shippers nominate gas in the day-ahead market, and may update their nominations at various points during the day in which the natural gas flows (see Pipeline Capacity Scheduling graphic below).

Pipeline Usage or Load Factor

Load factor measures the use of a pipeline network. It is the average capacity used at a given point or segment relative to a measurement of maximum or peak available capacity. Customers with a 100 percent load factor use their maximum capacity every day; one with a 50 percent load factor uses its capacity only half the time.

Different types of customers use pipeline capacity differently. Historically, industrial customers have exhibited high load factors and residential customers that primarily rely upon seasonal gas to heat homes have had lower load factors.

Pipelines are accustomed to serving different demands, which can affect how much of their capacity is used at various times. For example, Kern River Gas Transmission has operated at around 93 percent of capacity since 2005, while Algonquin Gas Transmission's capacity factor is considerably less. Algonquin's pipeline is used more seasonally than Kern River's, reflecting the seasonal demand in the Northeast.

Park and Loan Service

Park and loan service (PAL) is a way for shippers to balance their takes of gas with their supply, by providing a short-term load-balancing service to help shippers meet their load. Using the PAL service, shippers can take less gas than scheduled, thus parking their excess supply in the pipeline at times when the demand is lower than anticipated.

If demand is higher than expected, shippers can adjust their take upward, in effect borrowing gas from the pipeline.

PAL characteristics include:

- Park and loan services typically generate low revenue and are offered with the lowest service level priority among all pipeline services.
- Rates are based on costs associated with providing the services, such as plant costs if the services are offered.
- Market centers, or hubs, routinely offer these services.
- Charges are usually commensurate with interruptible service rates.
- Pipelines earn minimal revenue from park and loan.

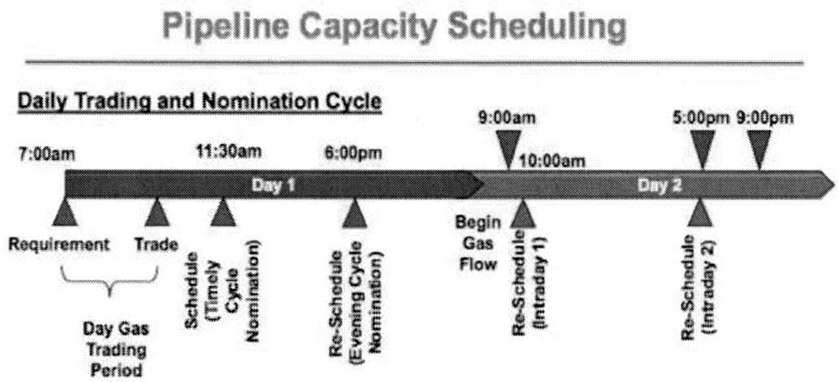

Cycle	Cycle Description
T	Timely Nominations sent by 11:30 AM and to the Pipeline by 11:45 AM to be effective at 9 AM next gas day
E	Evening Nominations sent by 6 PM and to the Pipeline by 6:15 PM to be effective 9 AM the next day. This is the cycle used for reporting on today's flow before the I2 cycle data is available.
I1	Intraday 1 nominations sent by 10 AM and to the Pipeline by 10:15 AM to be effective at 5PM the same day
I2	Intraday 2 nominations sent by 5 PM and to the Pipeline by 5:15 PM to be effective at 9 PM the same day. This is the cycle used for most Pipe2Pipe reporting purposes.

Pipeline Constraints and Capacity Growth

Pipeline capacity limits the supply that can be delivered to a specific region and is, therefore, a key factor in regional prices.

In recent years, the natural gas pipeline network has expanded significantly, removing bottlenecks and providing access to previously unreached supply areas.

A considerable amount of new pipeline capacity has been added in recent years to the Northeast. In 2008 and 2009, the region added 5.6 Bcfd in pipeline capacity. In 2010, it added another 1.2 Bcfd. Much of this new capacity was targeted at improving access to shale gas.

One of the largest additions to the natural gas infrastructure came with the completion of the 1.8-Bcfd Rockies Express Pipeline (REX) that moves natural gas from Wyoming to eastern Ohio. REX serves the dual role of relieving pipeline constraints that bottled up production in the Rockies and depressed prices there, while at the same time relieving constraints that increased prices in the East. Increased gas flows from the Rockies over REX, coupled with new shale supplies, have reduced prices in the Midcontinent and Northeast.

Meanwhile, Rockies producers saw a rise in prices. The Rockies gas flowing eastward displaced gas from the Permian Basin. The Permian gas, in turn, began moving to the Southern California market. Consequently, regional price differences moderated.

Other smaller projects had similar effects. New pipelines to increase the flow of Barnett Shale gas into the interstate network have had a secondary effect of reducing congestion across the Texas-Louisiana border.

The Florida Panhandle and Northern California used to be some of the most frequently constrained regions of the country, but each has received significant new pipeline capacity. Expansion of Florida Gas Transmission in 2011 added about 800 MMcfd, a boost of 33 percent, of gas transmission capacity to peninsular Florida. The 680-mile, 42-inch-diameter Ruby Pipeline, which began operations in 2011, now flows Rockies gas from Opal, Wyo., to Malin, Ore.

Local Distribution

Distribution lines typically take natural gas from the large interstate pipelines and deliver the gas to retail customers. While some large consumers – industrial and electric generators, for example – may take service directly off an interstate pipeline, most receive their natural gas through their LDC. LDCs typically purchase natural gas and ship it on behalf of their customers. They take possession of the natural gas from interstate pipelines at local citygates and deliver the natural gas to their customers at the customer's meter. According to the United States Department of Transportation's Pipeline and Hazardous Materials Safety Administration, this distribution involves a network of smaller pipelines totaling more than two million miles, as well as smaller scale compressors and meters.

Some states allow competition in natural gas service at the local level. In these circumstances, natural gas marketers purchase the natural gas and arrange for it to be shipped over both the interstate pipeline network and the local distribution pipeline system.

Storage

Natural gas production remains relatively unchanged day-to-day throughout the year, and recently has been around 60 Bcfd. Demand, however, changes considerably with the seasons. Natural gas storage enables producers and purchasers to store gas during periods of relatively low demand – and low prices – then withdraw the gas during periods of relatively higher demand and prices.

Typically, the amount stored or withdrawn is the difference between demand and production. This differs from storage capacity – the maximum amount that can be stored at any point in time. Working gas storage capacity, as tracked by EIA, is more than 4,100 Bcf. Gas may be stored in underground facilities. Storage capacity adds flexibility to pipeline and distribution systems and helps moderate prices by providing an outlet for excess gas during periods of low demand. Storage facilities also provide a readily accessible supply in periods of high demand. Some natural gas can also be stored in the pipelines as linepack, in which more molecules of gas are held in a segment of pipeline under greater-than-normal pressure.

EIA's weekly storage report provides a low-resolution snapshot of the natural gas supply and demand balance. EIA releases its storage report at 10:30 a.m. on Thursdays. The price for natural gas futures can change dramatically within seconds when the report comes out. If the reported injection or withdrawal significantly differs from market expectations, the price for natural gas futures may rise or fall.

Storage Facilities

The bulk of the storage capacity in the United States is below ground; differing cost and operational characteristics affect how each facility is used:

- Deliverability rate is the rate at which inventory can be withdrawn. The faster the natural gas can be removed from storage, the more suitable the storage facility is to helping serve rapidly changing demand.
- Cycling capability is the ability of the resource to quickly allow injections and withdrawals, which is useful for balancing supply and demand. Salt caverns tend to have high withdrawal and injection rates, enabling them to handle as many as a dozen withdrawal and injection cycles each year. LNG storage also demonstrates these capabilities.

Natural gas in an underground storage facility is divided into two general categories, working gas and base gas. Base gas is the volume of natural gas, including native gas, needed as a permanent inventory in a storage reservoir to maintain adequate reservoir pressure and deliverability rates throughout the withdrawal season. Working gas is the volume of gas in the reservoir above the designed level of base gas and that can be extracted during the normal operation of the storage facility.

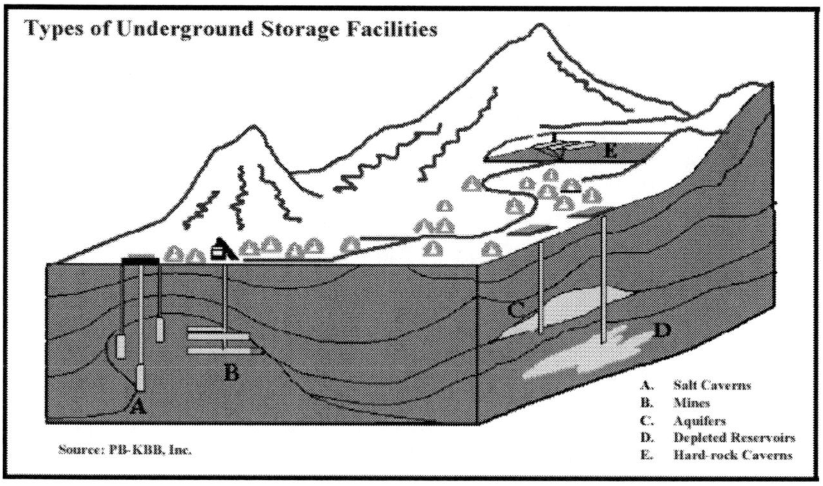

Most of the nation's gas storage is in depleted reservoirs (former oil and gas fields). These facilities reuse the infrastructure – wells, gathering systems and pipeline connections – originally created to support the field when it was producing. About 50 percent of total capacity goes to base gas used to maintain operating pressure at the facility, and inventory usually turns over once or twice a year.

Other storage facilities reside in aquifers that have been transformed into gas storage facilities. These are mostly in the Midwest. These aquifers consist of water-bearing sedimentary rock overlaid by an impermeable cap rock. Aquifers are the most expensive type of natural gas facility because they do not have the same retention capability as depleted reservoirs. Therefore, base gas can be well over 50 percent of the total gas volume. This makes the facility more sensitive to withdrawal and injection patterns, so inventory usually turns over just once a year. Salt cavern formations exist primarily in the Gulf Coast region. These air- and water-tight caverns are created by removing salt through solution-mining, leaving a cavern that acts as a pressurized vessel. Little base gas is required, which allows inventory to turn over as many as a dozen times during the year, and results in high injection and withdrawal rates. This flexibility has attracted new development, resulting in salt cavern storage growth over the past decade. Salt caverns generally hold smaller volumes than do depleted-reservoir or aquifer gas storage facilities. Natural gas may also be stored in above-ground tanks as LNG. There is LNG storage at all of the onshore LNG-receiving terminals, and there are about a hundred standalone LNG storage facilities in the United States, as well. Occasionally, LNG ships also provide storage. LNG storage is highly flexible, allowing multiple inventory turns per year with high injection and withdrawal rates.

Regional Storage

For storage purposes, the EIA has divided the United States into three regions: producing, East and West. Just over half of the underground storage in the United States, 2,200 Bcf, sits in the East, near population centers. Much of this is in aquifers and depleted fields. Almost 1,300 Bcf sits in the producing region, which has not only depleted fields but also the greatest concentration of more-flexible salt cavern storage. The remaining 550 Bcf is in the West, primarily in depleted fields. Of this total working gas capacity of more than 4,100 Bcf, at the begining of winter there will be 3,900 Bcf or more of gas injected into storage, making it about 95 percent full as the winter heating season begins.

Storage Service and Uses

Approximately 120 entities – including interstate and intrastate pipeline companies, LDCs and independent storage service providers – operate the nearly 400 underground storage facilities active in the continental United States, the EIA says. Facilities operated by interstate pipelines and many others are operated on an open-access basis, with much of the working gas capacity available for lease on a nondiscriminatory basis.

The ability to store large quantities of natural gas improves reliability and usually has a moderating influence on natural gas prices. Storage inventory augments gas supply during the winter, but acts as an additional demand component during the summer injection season. The storage injection season typically starts April 1 and continues through Oct. 31, when demand for gas heating is lowest. Storage withdrawals generally start in November and last throughout the winter.

The ability to use storage to provide for winter peaks creates an intrinsic storage value. This is the value from buying during cheaper periods of the year for use during higher-cost seasons. Depleted reservoirs or aquifers – with limited ability to turn over inventory – support this type of use. Local distribution companies or pipelines store their gas in these facilities to ensure adequate supplies for peak seasons, balance load and diversify their resources.

Storage may be priced at cost-based or market-based rates. Pricing mechanisms for low-cycling depleted fields and aquifers may use a traditional cost-of-service structure, including:

- capacity charges for firm contract rights to physical storage capacity;
- deliverability charges for transportation to and from the storage facility;
- withdrawal charges for the removal of gas from storage; and
- injection charges for the injection of gas into storage.

A salt cavern, with its ability to turn over inventory frequently and quickly, allows for additional uses, enabling users to capture extrinsic value. Many salt dome facilities can cycle between injection and withdrawal at almost a moment's notice, giving users greater flexibility. Entities leasing storage capacity may move gas in and out of storage as prices change in attempts to maximize profits or minimize costs. Storage may be a component in producer or consumer hedging strategies, helping them to manage the risk of price movements. Further, storage helps shippers avoid system imbalances and associated penalties, and supports swing gas supply services, which are short-term contracts that provide flexibility when either the supply of gas from the seller, or the demand for gas from the buyer, are unpredictable. Storage also facilitates title transfers and parking and lending services – interruptible service by which the customer injects or withdraws gas for a short period of time, usually a month. This helps shippers balance daily receipts and deliveries, manage their overall supply portfolio or take advantage of price movements. Consequently, storage operators have begun offering a more varied menu of services, and users have begun using storage as a commercial tool and as part of a comprehensive supply portfolio strategy.

Merchant storage, frequently using salt caverns, uses market-based prices, recognizing the dynamics affecting value at any given point in time. Prices often take into account the prices at which the Nymex futures contracts are trading. They may also reflect the storage volume, the number of times the gas will be cycled, the length of the contract and the timeframe it covers and the maximum daily quantity that may be injected or withdrawn. Energy marketers have increasingly used these facilities as they try to profit from price volatility. It is also attractive to shippers, industrial consumers with uncertain loads and gas-fired generators whose needs change rapidly.

Pipelines also offer storage service, both firm and interruptible, as part of their open access transportation service under FERC rules. Rates are rarely market-based. Instead, prices are based on cost of service, with rates containing reservation and usage components for firm service and a usage component for interruptible.

Market Effects

Storage can mitigate large seasonal price swings by absorbing natural gas during low demand periods and making it available when demand rises.

Further, storage levels can affect the market's expectations about prices during the coming winter high-demand season. The amount of gas in storage in November is a key benchmark of the gas industry's ability to respond to changes in winter weather. Higher storage levels tend to reduce forward prices; lower storage levels tend to increase them, all other market conditions being equal.

FERC Jurisdiction

The underground storage of natural gas has historically been critical in assuring that the needs of natural gas customers are met. The Energy Policy Act of 2005 added a new section to the Natural Gas Act stating that the Commission may authorize natural gas companies to provide storage and storage-related services at market-based rates for new storage capacity, even though the company cannot demonstrate it lacks market power. To make this authorization, the FERC must determine that market-based rates are in the public interest and are needed to encourage the construction of the capacity and that customers are adequately protected.

Natural Gas Pricing and Trading

The natural gas industry in the United States is highly competitive, with thousands of producers, consumers and intermediate marketers. Some producers have the ability to market their natural gas and may sell it directly to local distribution companies (LDCs), to large industrial buyers and to power plants. Other producers sell their gas to marketers who aggregate natural gas into quantities that fit the needs of different types of buyers and then transport the gas to their buyers.

Most residential and commercial customers purchase natural gas from a LDC. In contrast, many industrial customers and most power plants have the option to purchase natural gas from a marketer or producer instead of from the LDC, thereby avoiding any LDC charges.

Interstate pipelines do not buy and sell natural gas and are limited to providing transportation services only, including storage. As noted, interstate pipelines transport natural gas at rates approved by the FERC.

Natural Gas Marketers

Most gas trading in the United States is performed by natural gas marketers. Any party engaging in the sale of natural gas can be termed a marketer; however, marketers are usually specialized business entities dedicated solely to transacting in the physical and financial energy markets. It is commonplace for natural gas marketers to be active in a number of energy markets, taking advantage of their knowledge of these markets to diversify their business.

Marketers can be producers of natural gas, pipeline marketing affiliates, LDC marketing affiliates, independent marketers or large-volume users of natural gas. Some marketing companies may offer a full range of services, marketing numerous forms of energy and financial products, while others may be more limited in their scope. For instance, most marketing firms affiliated with producers do not sell natural gas from third parties; they are more concerned with selling their own production and hedging to protect their profit margin from these sales.

There are five classifications of marketing companies: major nationally integrated marketers, producer marketers, small geographically focused marketers, aggregators and brokers.

The major nationally integrated marketers offer a full range of services, and market numerous different products. They operate on a nationwide basis, and have large amounts of capital to support their trading and marketing operations. Producer marketers are those entities generally concerned with selling their own natural gas production or the production of their affiliated natural gas production company. Smaller marketers target particular geographic areas and specific natural gas markets. Many marketing entities affiliated with LDCs are of this type, focusing on marketing gas for the geographic area in which their affiliated distributor operates. Aggregators generally gather small volumes from various sources, combine them and sell the larger volumes for more favorable prices and terms than would be possible selling the smaller volumes separately. Brokers are a unique class of marketers in that they never take ownership of natural gas themselves. They simply act as facilitators, bringing buyers and sellers of natural gas together.

All marketing companies must have, in addition to the core trading group, significant backroom operations. These support staff are responsible for coordinating everything related to the sale and purchase of physical and financial natural gas, including arranging transportation and storage, posting completed transactions, billing, accounting and any other activity that is required to complete the purchases and sales arranged by the traders.

In addition to the traders and backroom staff, marketing companies typically have extensive risk-management operations. The risk-management team is responsible for ensuring that the traders do not expose the marketing company to excessive risk.

Market Hubs

Natural gas is priced and traded at different locations throughout the country. These locations, referred to as market hubs, exist across the country and are located at the intersection of major pipeline systems. There are more than 30 major market hubs in the United States, the principal one of which is known as the Henry Hub, located at Erath, in southern Louisiana. The price at which natural gas trades differs across the major hubs, depending on the supply and demand for natural gas at that particular point. The difference between the Henry Hub price and another hub is called the location differential, or basis. In addition to market hubs, other major pricing locations include citygates. Citygates are the locations at which distribution companies receive gas from a pipeline. Citygates at major metropolitan centers can offer another point at which natural gas is priced.

The most important market hub and pricing point in the United States is the Henry Hub, which is run by Sabine Pipe Line LLC and is a major intersection of pipelines. It has 12 delivery points and 4 major receipt points, all at that one hub, making it a crossroads for significant amounts of gas moving to locations across the country. The Henry Hub is also the delivery point for the New York Mercantile Exchange (Nymex) natural gas futures contract. Changes in price at Henry Hub provide a good indicator of how prices are generally changing across the country.

Basis usually reflects the variable cost to transport gas between Henry and another hub. Basis can change, sometimes dramatically, depending on local market conditions, and can widen considerably when pipelines between two points are congested. Basis in excess of transportation costs results from pipeline constraints and lack of pipeline competition. The gas price at a hub in Florida, for example, would be the price at Henry Hub and the basis to the Florida hub.

Physical Trading

Physical trading contracts are negotiated between buyers and sellers. There are many types of physical trading contracts, but most share some standard specifi cations, including specifying the buyer and seller, the price, the amount of natural gas to be sold (usually expressed in a volume per day), the receipt and delivery point, the tenure of the contract (usually expressed in number of days beginning on a specifi ed day) and other terms and conditions. The special terms and conditions usually outline such things as the payment dates, quality specifi cations for the natural gas to be sold and any other specifi cations agreed to by both parties.

Physical contracts are negotiated between buyers and sellers over the phone or executed on electronic bulletin boards and e-commerce trading sites.

There are three main types of physical trading contracts: swing contracts, baseload contracts and firm contracts:

- Swing (or interruptible) contracts are usually short-term contracts, and can be as short as one day and are usually not longer than a month. These contracts are the most flexible, and are usually put in place when either the supply of gas from the seller, or the demand for gas from the buyer, are unreliable.
- Baseload contracts are similar to swing contracts. Neither the buyer nor seller is obligated to deliver or receive the exact volume specified. However, it is agreed that both parties will attempt to deliver or receive the specified volume, on a best-efforts basis.
- Firm contracts are different from swing and baseload contracts in that both parties are legally obligated to either receive or deliver the amount of gas specified in the contract. These contracts are used primarily when both the supply and demand for the specified amount of natural gas are unlikely to change.

Price Discovery

Spot (Cash) Market

The U.S. natural gas marketplace has a highly competitive spot, or cash, market where brokers and others buy and sell natural gas daily. The daily spot market for natural gas is active, and trading can occur 24 hours a day, seven days a week. The map on the next page shows some of the points where natural gas for next-day physical delivery is actively traded on the IntercontinentalExchange (ICE). Some of these points are market centers, where brokers actively trade and prices are established. In addition to these market centers, natural gas is actively traded at many other locations, including segments of individual pipelines and locations where pipelines interconnect with LDCs.

Spot market transactions are normally conducted on electronic exchanges or by telephone, with the buyer agreeing to pay a negotiated price for the natural gas to be delivered by the seller at a specified delivery point. Natural gas spot prices reflect daily supply and demand balances and can be volatile.

Bidweek

The largest volume of trading occurs in the last five business days of a month, known as bidweek. This is the week when producers are trying to sell their core production and consumers are trying to buy for their core natural gas needs for the upcoming month. The core natural gas supply or demand is not expected to change; producers know they will have that much natural gas over the next month, and consumers know that they will require that much natural gas over the next month. The average prices set during bid-week are commonly the prices used in physical contracts.

Index Prices

Frequently the prices in longer-term contracts are indexed to prices that are regularly published in the trade press. Several publications, such as *Gas Daily, Natural Gas Intelligence* and *Natural Gas Week*, survey the market for daily transaction prices that are used to form and publish a daily index that is made available the night before or the morning of the next business day. Many market participants also report their bidweek prices to publications, which convert these prices into monthly locational price indexes that are available on the first business day following the last day of bidweek. These daily and monthly indexes, in turn, are used as the basis for pricing for those firms that do not choose to enter into fixed-price contracts (or are prohibited from using them by state or local regulators).

The Financial Market

In addition to trading physical natural gas, there is a significant market for natural gas derivatives and financial instruments in the United States. In the financial market, market participants are interested in profiting from the movement of the price of natural gas rather than delivering or receiving natural gas. The pricing and settlement of these financial products are tied to physical natural gas. It is estimated that the value of trading that occurs on the financial market is at least a dozen times greater than the value of physical natural gas trading.

Derivatives are financial instruments that derive their value from an underlying fundamental – in this case, the price of natural gas. Derivatives can range from being quite simple to being exceedingly complex. Traditionally, most derivatives are traded on the over-the-counter (OTC) market, which is essentially a group of market players interested in exchanging certain derivatives among themselves.

More information on financial markets appears in Chapter 4.

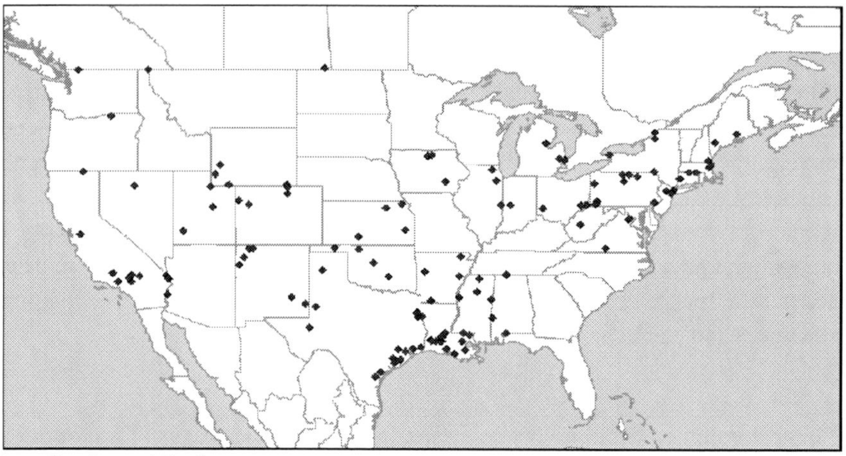

Source: InterContinental Exchange.

130 Hubs for Physical Trading on ICE.

3. Wholesale Electric Markets

Overview

Electricity is a physical product – the flow of electrons. It is a secondary energy source in that it results from the conversion of other energy forms such as natural gas, coal or uranium, or the energy inherent in wind, sunshine or the flow of water in a river. It may not be visible, but it can be turned on and off and measured.

Quick Facts: Measuring Electricity

Electricity is measured in terms of watts, typically in kilowatts (1,000 watts) or megawatts (1,000 kilowatts).

A kilowatt (or watt or megawatt) is the amount of energy used, generated or transmitted at a point in time. The aggregation of kilowatts possible at a point in time for a power plant, for example, is its capacity. The aggregation of kilowatts used at a point of time is the demand at that point. The number of kilowatts used in an hour (kilowatt-hour or kWh) is the amount of electricity a customer uses or a power plant generates over a period of time. Kilowatt-hours are used to bill customers.

Electric markets have retail and wholesale components. Retail markets involve the sales of electricity to consumers; wholesale markets typically involve the sales of electricity among electric utilities and electricity traders before it is eventually sold to consumers. This paper focuses on wholesale markets, although it addresses retail demand and other instances where retail markets strongly influence wholesale markets.

Much of the wholesale market and certain retail markets are competitive, with prices set competitively. Other prices are set based on the service provider's cost of service. For wholesale markets, FERC either authorizes jurisdictional entities to sell at market-based rates or reviews and authorizes cost-based rates.

In competitive markets, prices reflect the factors driving supply and demand – the physical fundamentals. In markets where rates are set based on costs, these fundamentals matter as well. Supply incorporates generation and transmission, which must be adequate to meet all customers demand simultaneously, instantaneously and reliably.

Consequently, key supply factors affecting prices include fuel prices, capital costs, transmission capacity and constraints and the operating characteristics of power plants. Sharp changes in demand, as well as extremely high levels of demand, affect prices as well, especially if less-efficient, more-expensive power plants must be turned on to serve load.

Background

Electricity on Demand

In the United States and other developed countries, consumers expect electricity to be available whenever they need it. Electricity use has grown enormously as consumers now consider not only refrigerators, TVs and hair dryers but also computers, iPods and other electronic devices as necessities. Consumers also expect to pay reasonable prices for the electricity they use.

Meeting these customer expectations is challenging. With few exceptions, electricity cannot be stored, in any appreciable quantities, and thus must be produced as needed. Further, unlike most other markets, electricity's historical inelastic demand does not move with prices. To provide electricity on demand, electric system operations have to be planned and conducted with that goal in mind. Lacking storage and responsive demand, operators must plan and operate power plants and the transmission grid so that demand and supply exactly match, every moment of the day, every day of the year, in every location.

The Drive for Enhanced Value

The electric industry has met this growing demand with increasing efficiency. Between 1929 and 1967, the national average cost of electricity for residential customers plummeted from about 60¢/kWh to 10¢/kWh (in 2005 dollars), and remains around there today. How did the industry achieve such tremendous cost savings and then keep the real price of electricity flat over the past 40 years? Part can be explained by greater efficiency – power plants use less fuel, and new techniques make it cheaper to extract the coal and natural gas that fuels generators. Another part of the answer, though, stems from changes in the way the industry is organized and operated.

Economies of Scale

Electric power is one of the most capital intensive industries. Generation alone can account for roughly 65% of a customer's electric bill. Spreading these relatively fixed costs over more customers helps bring down the cost that each customer pays.

Thomas Edison's first street lighting project in the 1880s grew to electrifying whole neighborhoods, towns and cities. Providing service over larger areas allowed utilities economies of scale in generating technology. The cost per unit of production dropped as power plants grew larger and larger. The companies building these facilities were basically self-contained – they owned and operated the generation, transmission and distribution facilities. Power lines were built from their generation to their population, or load, centers. These companies were vertically integrated.

One downside of larger generating units is that they are difficult to replace if they experience unexpected shutdowns. For a single utility building a new and larger unit, the only way to ensure reliable service is to build two units – creating a capacity reserve. When coal and nuclear unit sizes grew to 500 or 1,000 MW, building two units became very expensive for any individual company.

Reserve Sharing, Interconnection and Power Pools

The solution to high reserve costs was to share reserves with adjacent utilities. Instead of building two large units, utilities could buy from their neighbors in times of need, and cut their costs significantly.

To facilitate reserve sharing, utilities built major interconnecting transmission lines large enough to deliver power in case of a major generator outage. Today's bulk power grid began as a way to maintain reliable service while lowering costs.

As more utilities share reserves, the smaller the reserves each must carry, and the lower the costs. The value of reserve-sharing agreements led to the formation of power pools, the forerunners of today's regional transmission organizations (RTOs).

Coordinating reserves also led to closer coordination of other utility functions, such as the process of determining which generating units to use, called unit commitment. Operators want to commit just enough capacity for the next day to ensure reliability but no more than needed, to minimize costs. This began a new phase of using economies of scale in system operations encompassing whole regions of the country.

Regional coordination also was spurred by special circumstances, particularly in the West. Large federally owned dams on the Columbia and Colorado rivers generate power from the spring runoff of melting mountain snow. When the reservoirs are full and the turbines are spinning, there is not enough local demand to use the power.

Since the hydropower was cheaper than any alternative, long distance transmission lines were built to deliver the excess power from the Northwest and Southwest to load centers in California.

With the transmission interconnections in place, northwestern utilities found that they could get cheaper power from southern power generation at other times of the year. These seasonal and regional disparities in availability and price provide for a lively bilateral trading market.

In the 1960s, the electric industry created an informal, voluntary organization of operating staff to aid in coordinating the bulk power system. Then, in 1965, the largest power blackout until that time hit the northeastern United States – including New York – and southeastern Ontario, Canada, affecting 30 million people. The blackout led to the development in 1968 of the National Electric Reliabil ity Council (NERC), shortly thereafter renamed the North American Electric Reliability Council. and nine regional reliability councils. Rather than serving as a pool or other entity for sharing resources, NERC focused on reliability.

In 2006, using authority granted in the U.S. Energy Policy Act of 2005, FERC certified NERC as the electric reliability organization for the United States, and reliability standards became mandatory and enforceable.

Optimizing Unit Commitment and Economic Dispatch

The industry also reduced costs by using computers and communication technology to optimize system operations. Utilities use algorithms for optimizing the commitment of their generating units, while RTOs' day-ahead market software does this for suppliers bidding into their markets.

In real time, demand is changing all the time. Without storage and responsive demand, the output of some generators must change to follow constantly changing demand. This is known as load following. Utilities use economic dispatch to optimize the use of these units and minimize real-time costs.

Economy Energy Trade

Since transmission interconnections were built primarily for the rare need to deliver reserves in emergencies, the industry had excess transmission capacity. This allowed utilities to use the lines to trade power. Major utilities generally owned sufficient capacity to meet their own peak power needs. However, sometimes the cost of operating their marginal generation was higher or lower than that of their neighbors. Transmission availability provided opportunities for utilities to save money by buying energy when it was cheaper than generating and selling energy to utilities with higher costs. This is called economy energy trading.

Evolving Public Policies

Different public policy theories have shaped the electric power industry over its history. All of these public policies are still in play to some extent today. Five concepts that helped shape the electricity industry and markets are outlined below.

Not-for-Profit Utilities

One of the first approaches to ensuring customer value was to depend on nonprofit electric providers. In the early days of the industry, electrification started in towns and cities. In many places, this utility service was provided by the municipal government. The federal government stepped in to develop and market the nation's significant hydroelectric resources. The Depression-era rural electrification program relied on customer-owned rural electric cooperatives and low-interest government loans. There are currently more than 1,700 municipal and almost 900 cooperative utilities in the United States.

Regulated Natural Monopolies

A second model for operating power systems was investor-owned regulated monopolies. In the early days of the industry, while many cities went the municipal route, many investor-owned utilities were also starting up. These private utilities are regulated, typically by a state agency. Initially, they agreed to be regulated to overcome a lack of retail competition, and were granted exclusive service territories (franchise). Today, regulation focuses on mitigating market power, among other things, because many utility functions are seen as natural monopolies.

State regulators approve a utility's investments in generation and distribution facilities, either in advance of construction or afterwards when the utility seeks to include a facility's costs in retail rates. Some states eventually developed elaborate integrated resource planning (IRP) processes to determine what facilities should be built.

Power Pools

Power pools are multilateral arrangements with members ceding operational control over their generating units and transmission facilities to a common operator. Members provided incremental cost data about their units and system status data to the operator. The operator ran an energy management system that used the unit cost data to optimize on a multilateral basis unit commitment and economic dispatch.

PJM began in 1927 for utilities to share their generating resources, forming the world's first power pool. The New York Power Pool was formed in 1966 and the New England Power Pool in 1971 in response to the 1965 Northeast blackout. The Electric Reliability Council of Texas (ERCOT) and the Southwest Power Pool (SPP) formed in 1941 to pool resources for the war effort.

Competition, Part 1: Competitive Generation and Open Access

The environmental movement and initiatives to open the airline and trucking industries to competition also helped shape the energy industry in the 1970s. A provision in President Carter's energy plan led to passage of the Public Utility Regulatory Policies Act of 1978 (PURPA), which ushered in the next era.

PURPA established a program implemented by states and overseen by the FERC to encourage the use of efficient cogeneration (using the heat from industrial or other processes to generate electricity) and small scale renewable generation. FERC's role was to issue regulations for the program and certify that qualifying facilities (QFs) met statutory requirements. States administratively set the price to be paid to these generators at the cost the utilities would avoid by purchasing the power rather than generating it themselves.

Most states set their avoided cost rate so low that they got little QF capacity. However, California, Texas and Massachusetts set very generous avoided cost rates and were overwhelmed with QF capacity, much of which received prices that turned out to be higher than the actual costs avoided by the purchasing utility. The rapid growth and size of the QF industry surprised many policymakers and entrepreneurs, and got them thinking about the viability of generation independent of regulated monopolies.

In 1988, FERC proposed rules to allow states to set their avoided-cost rate based on an auction. Instead of taking all capacity at a set rate, states could set the rate based on bids to supply a certain amount of needed capacity. The Commission also proposed to open the avoided-cost auction up to independent power producers (IPPs) that did not qualify as QFs. In this way, a regulatory program was transformed into a competitive initiative.

Under the regulated monopoly model, utilities owning and operating transmission lines had no obligation to allow others to use them. This posed a significant barrier to the development of an independent power industry. The Commission started conditioning approval in merger cases with the voluntary provision of open transmission access. The Energy Policy Act of 1992 gave the Commission authority to grant transmission access on request. These approaches to open access resulted in patchwork transmission access.

By the mid-1990s, support for opening the transmission grid to all users encouraged the Commission to pursue a generic solution. Order 888 required mandatory open transmission access by all transmitting utilities and a reciprocity provision successfully extended open access to nonjurisdictional entities (municipal, cooperative and federal utilities).

Order 889 addressed matters needed to implement open access. The rule established the Internet-based Open Access Same-Time Information System (OASIS) for posting available transmission capacity and reserving transmission capacity. These rules required significant changes to utility control room operations and limited the ability of companies to share transmission-related information with their own power marketing operating units.

Competition, Part 2: Integrating Markets and Operations – RTOs

Order Nos. 888 and 889 were designed for an industry of bilateral energy markets, in which parties negotiated transactions among themselves. The open-access transmission tariff and rules did not work well for multilateral power pools open to independent power producers. This led to the development of independent system operators (ISO) and, subsequently, regional transmission organizations (RTO). This primer uses RTO to stand for both RTOs and ISOs.

RTOs did more than operate the transmission system and dispatch generation, however. They developed markets in which buyers and sellers could bid for or offer generation. The RTOs used the bid-based markets to determine economic dispatch.

Major parts of the country operate under more traditional market structures, notably the West (excluding California) and the Southeast. Notably, two-thirds of the nation's electricity load is served in RTO regions.

Electricity Demand

Americans use electricity for heat and light, to run machinery and to power a growing number of products such as televisions, radios, computers, hair dryers, cell phones and iPods.

This use has been increasing, reaching 3,865 gigawatt-hours (GWh) of electricity in 2008. Demand dropped in 2009 with the recession, but has since regained its upward trend.

The bulk of the electricity generated is sold to consumers, known as end-users or retail customers. Some consumers generate some or all of the power they consume. Some of the electricity sold to retail consumers is generated by integrated investor-owned utilities, federal entities, municipally owned and co-operatively owned utilities that sell the power directly to consumers. The rest of the electricity ultimately consumed by retail customers is bought and sold through wholesale electric markets.

This primer focuses on wholesale markets, which generally involve the sale of electricity to entities that resell the power to retail customers. However, retail consumers' electric use shapes demand and, therefore, the wholesale markets.

Quick Facts: Measuring Electricity

Electric use is described in terms of quantity and time.

The unit of measure of the quantity used is the kilowatt (kW), or 1,000 watts. The maximum number of all the kilowatts used by consumers on an electrical system at a point of time is peak demand.

The amount of electricity a consumer uses over a period of time is described as the number of the kilowatt-hour (kWh) - 1,000 watts working for one hour. Consumers pay based on the number of kWh they consume in a billing period, typically a month.

Source: EIA.

Demand Characteristics

Demand is often characterized as baseload or peak. Base-load is demand that occurs throughout the day or throughout the year. Refrigerators, for example, may create base-load demand. Peak load is demand that shows up during part of the day or year, all at the same time – heating or air conditioning, for example.

Demand for electricity follows cycles, throughout the day and year. Regionally, electric demand may peak in either the summer or the winter. Spring and fall are typically shoulder months, with lower peak demand. Seasonal peaks vary regionally, although the highest levels of power load in almost all regions of the United States occur during heat waves and are most acute during the daily peak load hours reached in the late afternoon. However, a minority of regions reach their peak load when the weather is extremely cold. These are primarily areas with significant space-heating requirements and little summer air conditioning load. A majority of these systems are in the far northern areas of the United States, where air conditioning load is not significant. South Florida's seasonal peak also occurs during the winter, when the population and tourism surges and uses more power than native Floridians do in the summer.

Daily demand typically peaks in the late afternoon, as commercial and domestic activities peak, and, in the winter, when lighting needs grow.

Electricity use also varies between weekdays and weekends. Commercial and industrial activities are lower on weekends and peoples' noncommercial activities change with their personal schedules. The load on different weekdays also can behave differently. For example,

Mondays and Fridays, being adjacent to weekends, may have different loads than Tuesday through Thursday. This is particularly true in the summer.

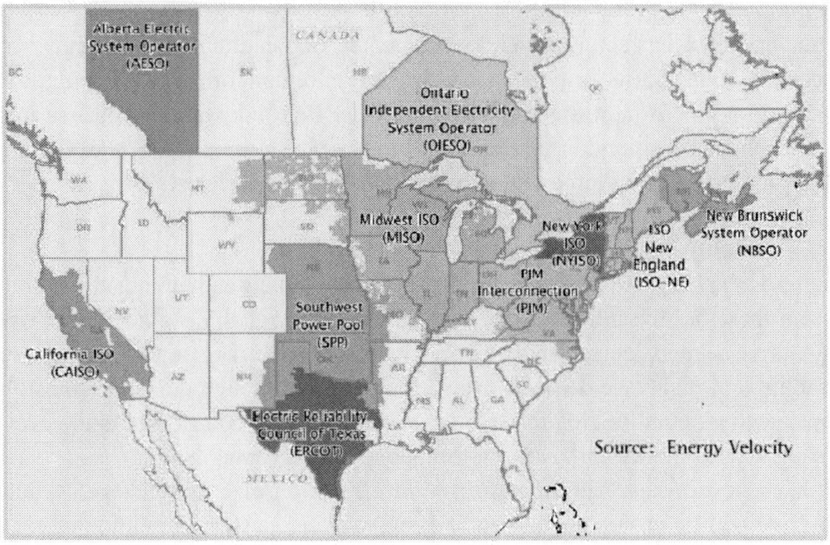

North American Regional Transmission Organizations.

Because demand historically has not varied with price and because storage options are limited, generation must rise and fall to provide exactly the amount of electricity customers need. The cost of providing power typically rises as demand grows, and falls as demand declines, so wholesale power prices are typically highest during peaks. Consequently, system planners, power marketers and traders all carefully track weather trends, economic growth and other factors to forecast power demand.

Demand Drivers

In general, the amount of electricity demanded is relatively insensitive to the price of electricity in the short-term (inelastic). One reason for this is that many customers – especially smaller customers – do not get price signals to which they can respond. Most residential customers are billed monthly on a preset rate structure. Large industrial customers, on the other hand, may receive real-time price signals.

Further, electricity is a necessity to most people and businesses. While they may be able to reduce their demand in the short-term – by turning down the thermostat or turning off lights, for example – electric consumers find it difficult to do without electricity altogether. There is little storage for electricity now and few realistic substitutes. Consequently, demand tends to drive price, especially when the system is stressed.

In the longer-term, options for reducing electricity use include switching to natural gas, installing insulation and implementing other energy efficiency measures. Larger consumers may consider building their own generation facilities.

Governments and businesses are also developing demand-response programs, which provide plans in which customers agree to reduce load in exchange for compensation.

Factors driving demand include demographics, climate and weather, economic activity and policies and regulations.

Demographics

Population levels affect demand, with greater population levels tending to increase electric consumption. Shifts in population also affect regional demand. Population flight in the 1980s from northern industrial regions – the Rust Belt – to warmer climates in the South affected residential consumption patterns. In the 1990s, consumption in the South surpassed that in the Midwest, making it the region with the greatest electric use.

Climate and Weather

Weather is the biggest factor driving demand. General climatic trends drive consumption patterns and therefore the infrastructure needed to ensure reliable service. Cold weather and short days drive winter demand in northern regions. Southern regions rely more on electric space heating, and, thus, see demand rise in the winter, although demand typically peaks in the summer with air conditioning load. In the winter, lighting contributes to the occurrence of peaks during the seasonally dark early morning and early evening hours.

Weather also can have extreme short-term effects on electricity usage. A sudden cold snap can drive heating use up quickly and a heat wave can push air conditioning loads. Other, less obvious weather patterns affect demand – rain and wind, for example, may result in sudden cooling, affecting heating or air conditioning.

Economic Activity

The pattern of socioeconomic life affects the cycle of electric use, with weekends and holidays showing a different pattern than weekdays. Demand typically rises as people wake up and go to work, peaking in the afternoon.

The health of the United States and regional economies also affects power demand. During periods of robust activity, loads increase. Similarly, loads drop during recessions. These changes are most evident in the industrial sector, where business and plants may close, downsize or eliminate factory shifts. In addition to reducing overall demand, these changes may affect the pattern of demand; for example, a factory may eliminate a night shift, cutting baseload use but continuing its use during peak hours. In some cases these effects can be significant.

Quick Facts: Heating and Cooling Degree Days

In the United States, engineers developed the concept of heating and cooling degree days to measure the effects of temperature on demand. Average daily temperatures are compared to a 65° F standard - those in excess of 65° yield cooling degree days; those below 65° yield heating degree days. A day with an average temperature of 66° would yield one cooling degree day.

Energy Policies and Regulations

State regulatory agencies set prices and policies affecting retail customer service. Some states are considering changes that would enable customers to receive more accurate price signals. They include, among other things, changing rate structures so that the rate varies with the time of day, or is even linked to the cost of providing electricity.

Efforts to reduce overall demand by improving energy efficiency are underway through several governmental and utility venues.

Retail Customer Mix

Most electric utilities serve different types of customers: residential, commercial and industrial. Each class uses electricity differently, resulting in a differing load profile, or the amount that each customer class uses and the daily shape of the load. If a consumer uses electricity consistently throughout the day and seasons, his load shape is flat, and the load will be baseload. Another consumer may use more at some times than others, resulting in baseload and peaks. Greater variability in demand is typically more expensive to serve, especially if the peak occurs at the same time other customers' use peaks. Consequently, the mix of customer types affects a region's overall demand.

Residential consumers form the largest customer segment in the United States at approximately 38 percent of electricty demand. Residential consumers use electricity for air conditioning, refrigerators, space and water hearing, lighting, washers and dryers, computers, televisions, cell phones and other appliances. Prices for residential service are typically highest, reflecting both their variable load shape and their service from lower-voltage distribution facilities, meaning that more power lines are needed to provide service to them.

Commercial use is the next largest customer segment at approximately 36 percent, and includes hotels and motels, restaurants, street lighting, retail stores and wholesale businesses and medical religious, educational and social facilities. More than half of commercial consumers' electric use is for heating and lighting.

Industrial consumers use about 26 percent of of the nation's electricity. This sector includes, for example, manufacturing, construction, mining, agriculture and forestry operations. Industrial customers often see the lowest rates, reflecting their relatively flat load structure and their ability to take service at higher voltage levels.

Transportation demand for electricity stems primarily from trains and urban transportation systems. This is less then 1 percent of electricty demand.

Load Forecasting

Demand is constantly changing, challenging grid operators and suppliers responsible for ensuring that supply will meet demand. Consequently, they expend considerable resources to forecast demand. Missed forecasts, where actual demand differs significantly from the forecast, can cause wholesale prices to be higher than they otherwise might have been.

Forecasts are necessary as well for the variety of actions that must occur if sufficient supply is to be available in the immediate or long term: planning the long-term infrastructure needs of the system, purchasing fuel and other supplies and staffing, for example. Load forecasts are also extremely important for suppliers, financial institutions and other participants in electric energy generation, transmission, distribution and trading.

Load forecasting uses mathematical models to predict demand across a region, such as a utility service territory or RTO footprint. Forecasts can be divided into three categories: short-term forecasts, which range from one hour to one week ahead; medium forecasts, usually a week to a year ahead; and long-term forecasts, which are longer than a year. It is possible to predict the next-day load with an accuracy of approximately 1%-3% of what actually happens.

The accuracy of these forecasts is limited by the accuracy of the weather forecasts used in their preparation and the uncertainties of human behavior. Similarly, it is impossible to predict the next year peak load with the similar accuracy because accurate long-term weather forecasts are not available.

The forecasts for different time horizons are important for different operations within a utility company. Short-term load forecasting can help to estimate transmission system power flows and to make decisions that can prevent overloading of transmission systems. Timely implementation of such decisions leads to the improvement of network reliability and to the reduced occurrences of equipment failures and blackouts. Forecasted weather parameters are the most important factors in short-term load forecasts; temperature and humidity are the most commonly used load predictors.

The medium- and long-term forecasts, while not precise, take into account historical load and weather data, the number of customers in different customer classes, appliances used in the area and their characteristics, economic and demographic data, and other factors. For the next-year peak forecast, it is possible to provide an estimated load based on historical weather observations.

Long-term forecasts are used for system infrastructure planning and are meant to ensure that there are sufficient resources available to meet the needs of the expected future peak demand. These forecasts are made for periods extending 10 to 20 years into the future.

Demand Response

Electric demand is generally insensitive to price, meaning that demand does not typically fall significantly when wholesale prices rise. However, some utilities and grid operators are developing ways to stimulate a response from consumers through demand-response programs.

Demand response (DR) is the ability of customers to respond to either reliability or price triggers by forgoing power use for short periods, by shifting some high energy use activities to other times or by using onsite generation. The programs may use price signals or incentives to prompt customers to reduce their loads. The signals to respond to electric power system needs or high market prices may come from a utility or other load-serving entity, a regional transmission organization (RTO) or an independent DR provider. These programs are administered by both retail and wholesale entities.

DR has the potential to lower systemwide power costs and assist in maintaining reliability. It can be used instead of running power plants or to relieve transmission congestion There can also be environmental benefits because peaking units tend to be costly - and dirty - to run.

Demand response rewards consumers for reducing load during certain market conditions and at specific times. However, it is difficult to measure and quantify this reduction. Measuring and verifying the reduction requires development of consumers' baseline usage, against which their actual use is measured to determine the reduction in the event they are called to lessen their load. An accurate measure of their typical usage is important to prevent (or detect) gaming by participants.

Demand-Response Programs

Programs generally fall into three categories: curtailing, shifting or on-site generation.

Curtailing, or forgoing, involves reducing power use (load) during times of high prices or threats to reliability without making up the use later. For example, residential customers might turn off lights or raise thermostats during hot weather. Commercial facilities may turn off office equipment, lower building lighting or change thermostat settings by a few degrees.

Shifting involves moving or rescheduling high energy-use activities in response to high prices or DR program events to off-peak periods – evenings, nights or weekends. Industrial customers might reschedule batch production processes to evening hours or the next day.

Commercial establishments may delay high-energy operations. Residential customers may wait until evening or night to use high-energy consuming appliances, such as clothes dryers or dishwashers. In shifting, the lost amenity or service is made up at a subsequent time.

On-site generation is when some customers may respond by turning on an on-site or backup emergency generator to supply some or all of their electricity needs. Although customers may have little or no interruption to their electrical usage, their net load and requirements from the power system are reduced. The ability to use on-site generation is most common for institutional customers, such as hospitals, large schools or data centers.

DR programs can be further distinguished by whether they are controlled by the system operator (dispatchable) or the customer (nondispatchable). Dispatchable demand response refers to programs that reduce customer energy use, such as direct load control of residential appliances or directed reductions to industrial customers.

Dispatchable DR is used for reliability or economic reasons. Nondispatchable demand response lets the retail customer decide whether and when to reduce consumption in response to the price of power. It includes time-sensitive pricing programs based on rates that charge higher prices during high-demand hours and lower prices at other times.

As a result of technology innovations and policy directions, new types and applications of DR are emerging that encompass the use of smart appliances that respond in near real-time to price or other signals. These models may allow customers to respond more easily as they require little customer monitoring or interaction.

Demand Response in Retail Markets

Many states require utilities to use energy efficiency, DR or renewable resources. Energy Efficiency Resource Standards (EERS) in more than half of the states (plus Washington, D.C.) require utilities to achieve electric energy savings; many of these standards include peak load reduction targets. These mandates provide incentives for utilities to reduce customers' energy consumption, such as mechanisms that decouple profits from the amount of electricity sold, or performance bonuses for utilities that meet or exceed reduction targets.

Some states are implementing dynamic pricing, in which retail rates change frequently to better reflect system costs. Time-based rates depend on advanced meters at customer premises that can record usage. In time-of-use programs, customers are charged different prices at different times, with hours of peak demand costing more than off-peak hours.

In real-time pricing (RTP) programs, customers are charged prices reflecting the immediate cost of power. Industrial or very large commercial customers are often on RTP tariffs.

Critical peak pricing (CPP) uses real-time prices at times of extreme system peak, and is restricted to a small number of hours annually but features prices higher than time-ofuse prices during the critical peak. Consumers do not know in advance when a critical peak might be called. A CPP program for residential customers uses a carrot without the stick: critical-peak rebates. Participating customers get rebates on their bills for responding to utility price-signals, but are not penalized if they do not lower use in those hours.

Wholesale Market Programs

Retail programs may aid RTOs, although the RTO may not be able to invoke them or even see specifically the amount of response that occurs. Wholesale-level DR occurs in the RTOs, which differ in how demand-response resources (DRR) may participate in their markets. Some RTOs permit DRR to participate in their markets as voluntary reliability resources. For example, NYISO has an emergency demand-response program, which permits DRR to participate through an aggregator or other interface party, and receive energy payments for providing curtailments when called.

DRR also can participate in wholesale markets as capacity resources and receive advance reservation payments in return for their commitment to participate when called. Resources that fail to perform when called are penalized.

Finally, DRR can bid into RTO day-ahead (DA) markets as energy resources, specifying the hours, number of megawatts and price at which they are willing to curtail. ISOs set minimum bid values – NYISO's program has a $75/ MWh floor. In New York, a resource scheduled in the DA market is obligated to curtail, and failure to perform results in a penalty.

Some of the RTO DR comes from individual entities; the rest is accumulated through third-party aggregators, or curtailment service providers (CSPs), who recruit customers too small to participate on their own, such as schools, commercial chains or groups of residential customers. In aggregating small customers, CSPs have increased customer participation in many wholesale reliability and emergency programs. In NYISO's two incentive-based programs, CSPs increased their share of subscribed DRR to 77 percent in 2008 from 44 percent in 2003. CSPs were responsible for more than 60 percent of total DRR capacity, and 70 percent of new DRR in ISO-NE.

Demand-Response Use in Planning and Operations

Different DR programs can be used at various times to support planning and operations (see graphic, page 50). Energy efficiency programs that reduce baseload or peak demand over the long-term are incorporated into system planning. Dispatchable programs that are quickly implemented and targeted for short-term peak reductions – such as direct load control – lie on the other end of the spectrum, and are used in the moment of operation.

Electricity Supply and Delivery

Unlike many other products, electricity cannot be stored in any appreciable quantities. Further, electricity is a necessity for most consumers, whose use responds little to price changes. Finally, electric equipment and appliances are tuned to a very specific standard of power, measured as voltage. Deviations in voltage can cause devices to operate poorly or may even damage them. Consequently, the supply side of the electric market must provide and deliver exactly the amount of power customers want at all times, at all locations. This requires constant monitoring of the grid and close coordination among industry participants.

Electricity service relies on a complex system of infrastructure that falls into two general categories: generation and the delivery services of transmission and distribution. Together, the power generation and high-voltage transmission lines that deliver power to distribution facilities constitute the bulk power system. Transmission and distribution facilities are also referred to as the power grid. These are coordinated and at times operated by a grid coordinator.

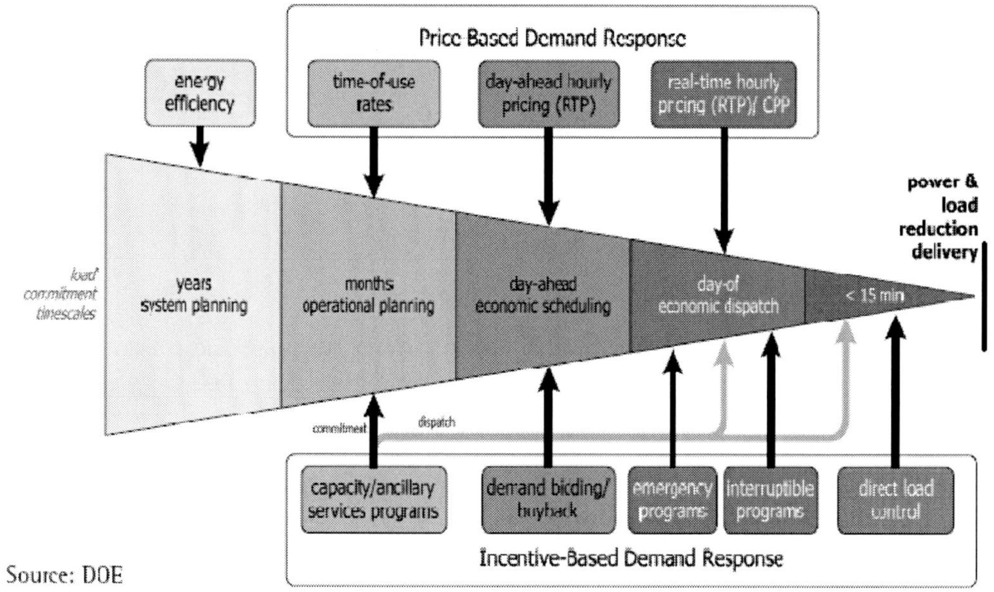

Demand-Response Program Use in Electric System Planning and Operations.

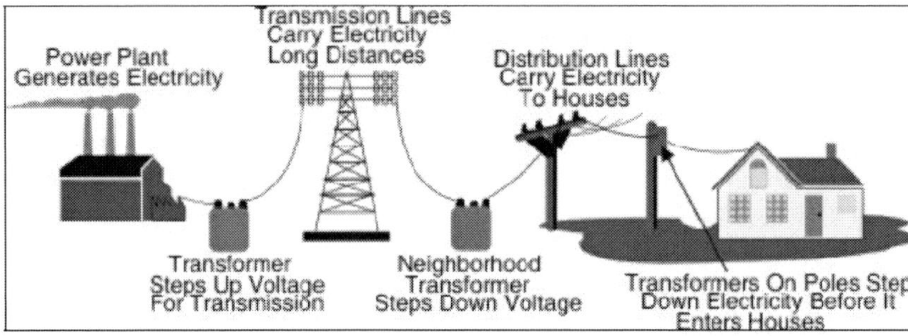

Source: National Energy Education Development Project.

Electricity Service Schematic.

Nationally, the grid is split into three main sections – the Western, Eastern and Texas Interconnections. These sections operate independently and have limited interconnections between them.

The nation, along with Canada and a small part of Mexico, is also divided into regional entities, (see map, page 51). The regional reliability entities fall under the purview of the North American Electric Reliability Corp. (NERC), which was designated by FERC as the nation's energy reliability organization and which develops standards, among other things, to ensure the grid's reliability. The standards, once issued by FERC, must be met by all industry participants – the standards are mandatory and enforceable. Consequently, the grid is designed and operated to meet these standards.

NERC's regions include:

- Florida Reliability Coordinating Council (FRCC),
- Midwest Reliability Organization (MRO),
- Northeast Power Coordinating Council (NPCC),
- Reliability First Corporation (RFC),
- SERC Reliability Corp. (SERC),
- Southwest Power Pool (SPP),
- Texas Reliability Entity (TRE) and
- Western Electricity Coordinating Council (WECC)

Generation

Power generators are typically categorized by the fuel they use and subcategorized by their specific operating technology. The United States has more than 1,000 gigawatts (GW) of total generating capacity. Coal, natural gas and nuclear dominate the power generation market.

Power plants each have differing costs and operational characteristics, both of which determine when, where and how plants will be built and operated.

Plant costs fall into two general categories: capital investment costs, which are amounts spent to build the plant, and operational costs, the amounts spent to maintain and run the plant. In general, there is a trade-off between these expenses: the most capital intensive plants are the cheapest to run – they have the lowest variable costs – and, conversely, the least capital intensive are more expensive to run – they have the highest variable cost. For example, nuclear plants produce vast amounts of power at low variable costs, but are quite expensive to build. Natural gas-fired combustion turbines are far less expensive to build, but are more expensive to run.

Grid operators dispatch plants – or, call them into service – with the simultaneous goals of providing reliable power at the lowest reasonable cost. Because various generation technologies have differing variable costs, plants are dispatched only when they are part of the most economic combination of plants needed to supply the customers on the grid. For plants operating in RTOs, this cost is determined by the price that generators offer. In other areas, it is determined by the marginal cost of the available generating plants.

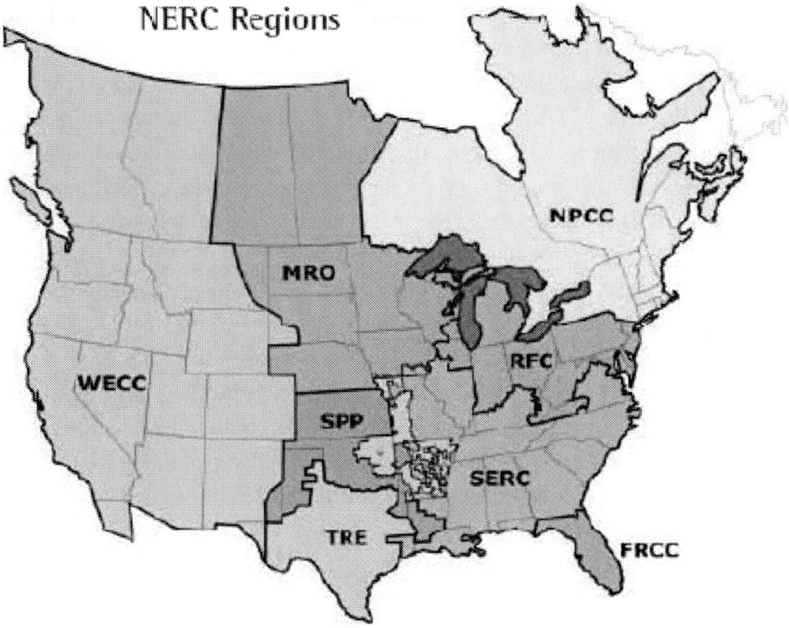

Construction of different generating technologies is subject to a number of issues, including community concerns, regional emission restrictions and the availability of fuels or other necessary resources:

- Wind plants are generally built in areas with the appropriate meteorological conditions. In most cases, these sites are located in rural areas with limited transmission access. For example, in West Texas, the transmission lines connecting wind farms with consumer centers in Dallas and Houston can become overloaded, requiring generators to curtail production.
- Coal plants have environmental characteristics that limit both their siting and operations. Specifically, they emit NO_x, SO_x, particulates, mercury and substantially higher levels of CO_2 than gas-fired plants. This has made financing these plants and siting them near urban centers difficult.
- There have been virtually no new nuclear plants built in the United States in the past 30 years. The technology of older plant designs became a source of concern following the accident at the Three Mile Island plant in the United States in 1977, the Chernobyl plant meltdown in Ukraine in 1986 and the Japanese earthquake, tsunami

and nuclear plant destruction in 2011. New plant designs have been put forward over the past few years and are expected to be very expensive and controversial to build. Further, the disposition of high-level radioactive waste remains an unresolved problem, and the waste remains at plant locations.

Conventional Generation

Natural gas power plants: These feature three major technologies, each with its distinct set of market advantages and limitations. They are steam boilers, gas turbines and combined cycle generators. Natural gas fuels nearly a third of electricity generation.

Steam boiler technology is an older design that burns gas in a large boiler furnace to generate steam at both high pressure and a high temperature. The steam is then run through a turbine that is attached to a generator, which spins and produces electricity. Typical plant size ranges from 300 MW to 1,000 MW. Because of their size and the limited flexibility that is inherent in the centralized boiler design, these plants require fairly long start-up times to become operational and are limited in their flexibility to produce power output beyond a certain range. Furthermore, these plants are not as economical or easy to site as newer designs – which explains why none has been built in recent years.

Gas turbines (GT) are small, quick-start units similar to an aircraft jet engine. These plants are also called simple cycle turbines or combustion turbines (CT). GTs are relatively inexpensive to build, but are expensive to operate because they are relatively inefficient, providing low power output for the amount of gas burned, and have high maintenance costs. They are not designed to run on a continuous basis and are used to serve the highest demand during peak periods, such as hot summer afternoons. GTs also run when there are systemwide shortages, such as when a power line or generator trips offline. GTs typically have a short operational life due to the wear-and-tear caused by cycling. The typical capacity of a GT is 10-50 MW and they are usually installed in banks of multiple units.

Combined cycle power plants (CCPPs) are a hybrid of the GT and steam boiler technologies. Specifically, this design incorporates a gas-combustion turbine unit along with an associated generator, and a heat recovery steam generator along with its own steam turbine. The result is a highly efficient power plant. They produce negligible amounts of SO_2 and particulate emissions and their NO_x and CO_2 emissions are significantly lower than a conventional coal plant. CCPPs, on average, require 80 percent less land than a coal-fired plant, typically 100 acres for a CCPP versus 500 acres for comparable coal plant, and CCPPs also use modest amounts of water, compared to other technologies.

Coal plants: These generate more than one-third of the electricity in the United States. These facilities tend to be large, baseload units that run continuously. They have high initial capital costs and are also somewhat complex in their design and operations. However, coal plants have low marginal costs and can produce substantial amounts of power. Most of the coal-fired plants in the United States are owned by traditional utility companies and located in the Southeast and Midwest.

Oil-fired plants: These play a minor role in U.S. power markets. These facilities are expensive to run and also emit more pollutants than gas plants. These plants are frequently uneconomic and typically run at low capacity factors. Like gas-fired generators, there are several types of units that burn oil; primarily, these are steam boilers and combustion turbines.

Generally, two types of oil are used for power generation: number 2 and number 6 (bunker) fuel oil. Number 2 is a lighter and cleaner fuel. It is more expensive, but because it produces fewer pollutants when burned, it is better for locations with stringent environmental regulations such as major metropolitan areas. Conversely, number 6 fuel oil is cheaper, but considered dirty because of its higher emissions. It is highly viscous (thick and heavy) and it comes from the bottom of the barrel in the refining process.

Nuclear plants. These provide roughly 20 percent of the nation's electricity; there are 104 operating plants with a total capacity of 100 GW. These plants are used as base-load units, meaning that they run continuously and are not especially flexible in raising or lowering their power output. Nuclear plants have high capital and fixed costs, but low variable costs, which includes fuel cost. They typically run at full power for 18 month, which is the duration of a unit's fuel cycle. At that point, they are taken off-line for refueling and maintenance. Outages typically last from 20 days to significantly longer, depending on the work needed.

Following the Three Mile Island plant accident in 1977, there was a cessation in the development of new plants. Most projects under construction in 1977 were finished, albeit with tremendous cost overruns. The last unit built in the United States came online in 1996.

Renewable Generation

Renewable resources use fuels that are not reduced or used up in the process of making electricity. They generally include biomass, geothermal, hydropower, solar, onshore and offshore wind, hydrokinetic projects, fuel cells using renewables and biogas.

Renewable generation provides a small percent of total U.S. capacity and generation. Even in 2009, when total U.S. electric output fell, average renewable generation grew 12% and wind output grew 28%, spurred by state regulations and federal tax credits. As renewable generation becomes a larger percentage of generation resources, integrating them into the operating power grid has presented challenges.

Capacity: Wind and solar capacity have grown faster than other renewable resources in recent years. Geothermal has more installed capacity than solar, but is growing more slowly.

Wind added the second highest amount of capacity after gas-fired generation in 2009 and 2010.

Additions are usually reported in megawatts of nameplate capacity. Actual capability varies from the nameplate for any unit type due to age, wear, maintenance or ambient conditions. But as renewable resources are often weather-dependent, their capacity factors – the ratio of average generation to the nameplate capacity for a specific period – have been much lower (as low as 30 percent) than for fossil-fuel-fired generation. Markets care about the difference between nameplate and capacity factor values when they evaluate capacity available to cover expected load. Prior to sufficient operating experience with a renewable technology, markets usually estimate capacity value conservatively.

Average capacity factors for new renewable resources in early 2010 were 34-35 percent for large wind projects, 74 percent for geothermal, 15-21 percent for commercial solar photovoltaic (PV), 35 percent for concentrating solar power (CSP) without storage (or 43-45 percent with storage technology) and 63 percent for biomass. Capacity factors have risen with technological innovation and improved manufacturing processes.

Characteristics: Wind power is the fastest-growing renewable resource, in part due to earlier cost declines and technology improvements as well as earlier receipt of federal tax credits. A 1.5-MW wind turbine was the most frequently installed size in 2009, although 2.3-MW turbines later became more common. Wind is largely pollution-free and can be located on farms and ranches.

Because the best wind resources are often far from load centers, insufficient transmission presents a challenge to delivering its output. Other market challenges for future wind development include its variable output, which is often inversely correlated to demand (seasonally and daily); system operators' inability to dispatch wind resources to meet load increases; difficulties related to accurately forecasting its ramping; and the need for companion generation (usually fossil-fueled) to be available to balance wind generation when the wind is not blowing.

Geothermal energy taps into reservoirs of steam and hot water deep beneath the earth's surface to produce power. The best resources are in the intermountain West. Geothermal potential is determined by thermal conductivity, thickness of sedimentary rock, geothermal gradient, heat flow and surface temperature. While geothermal power was less than 0.4 percent of U.S. generation in 2009, it was 11 percent of nonhydro renewable output in April 2010. California hosts more than 80 percent of U.S. operating capacity. The five states with the most geothermal capacity in development are California, Nevada, Oregon, Utah and Idaho.

Solar energy transforms sunlight into electricity using one of two technologies: photovoltaic (PV) or concentrating solar power (CSP). PV modules, or panels, transform sunlight directly into power using silicon wafers or nonsilicon thin-film technologies. They can be installed on roofs of buildings or at ground-level PV farms. CSP plants use a two-step process to transform the sun's energy. First, mirrors direct sunlight towards a receiver that captures the heat. CSP then employs a thermal process to create steam, driving an engine or turbine to produce electricity. CSP plants, which are dispatchable, can include low-cost energy storage that extends their availability later in peak hours. PV growth has been highly concentrated as a result of state policy incentives: 10 states have 95% of PV capacity; California alone has more than 50 percent. Annual PV additions rose to more than 100 MW beginning in 2006, spurred by tax incentives in the Energy Policy Act of 2005. More recent growth was spurred by falling costs, technology innovation, expanded federal tax benefits and an increase in state policies promoting investment. In 2010, 432 MW of CSP was operational and 81 MW was under construction. Seven western and southwestern states have extensive CSP potential: Utah, New Mexico, Arizona, Nevada, Texas, California and Colorado. Developing that potential will require overcoming challenges of siting, transmission and the need for extensive water supplies to clean mirrors.

Hydroelectric power is created when the kinetic energy of falling water drives turbine generators, which convert the energy into electricity. There are two types of hydroelectric projects: conventional and pumped storage. Conventional projects, which use a dam in a waterway, can operate in a run-of-river mode, in which water outflow from the project approximates inflow, or in a peaking mode, in which the reservoir is mostly drained to generate power during peak periods when energy is more valuable. Pumped storage projects use bodies of water at two different elevations. Water is pumped into elevated storage reservoirs during off-peak periods when pumping energy is cheaper; the water is then used to generate power during peak periods as it flows back to the lower elevation reservoir. Pumped storage is the only significant commercially deployed electricity storage technology available today.

Biomass includes many waste byproducts, such as agricultural residues, landfill gas, municipal solid waste and wood resources. The largest biomass category is wood waste, burned for heat and power in the lumber, pulp and paper industries. Challenges to biomass production include impacts on food supplies (for example, converting corn into ethanol), conserving natural resources and minimizing water pollution. State policies on renewable generation differ on eligibility of biomass technologies.

Biogas energy is created through the anaerobic (without oxygen) bacterial decomposition of manure, which is turned into a gas containing 60-70 percent methane. Biogas recovery can be installed at farms anywhere, used to run farm operations and reduce methane emissions from natural manure decomposition.

Renewable Energy Policies

Renewable development is frequently tied to policies promoting their use because of their higher cost relative to other technologies. Financial incentives include tax credits, low-cost loans, rebates or production incentives. Federal funding of research and development (R&D)

has played an important role in lowering costs or reducing the time it takes for renewable technologies to become commercially viable.

Congress has passed tax incentives to spur renewable resource investments. Wind, biomass, geothermal, marine and hydrokinetic project developers can receive federal production tax credits (PTC) based on a facility's production. It is an inflation-adjusted credit that runs for 10 years from the date a facility goes online. Initially set at 1.5¢/kilowatt hour (kWh), its value in 2010 was 2.1¢/kWh. To qualify, a facility has to be operational before the PTC expires.

Unlike other renewables projects, solar projects are eligible for a federal investment tax credit (ITC), worth 30 percent of a project's equipment and construction costs. The ITC begins the year a project starts commercial operation and depreciates over five years. The Emergency Economic Stabilization Act (2008), the legislation enacted to help shore up the U.S. financial system, extended the solar ITC for eight years, through 2016. It also expanded both the PTC and ITC to include utilities, which were previously ineligible for these credits. This change created the impetus for a model of utility owned and operated renewable generation, and led to a flurry of project announcements.

Provisions of the American Reinvestment and Recovery Act (ARRA) of 2009 extended the PTC and gave developers new options. It extended the credit for wind to 2012 and for other eligible technologies to 2013, and gave PTC-qualified developers the option to claim the 30 percent ITC on a project-by-project basis for the PTC's current duration.

Due to the economic crisis, ARRA gave developers another option for projects that began construction by the end of 2010 – they could apply for Treasury-administered cash grants, which monetized the ITC's value up front. ARRA funds helped support renewable energy research and development and aided capacity growth in 2009, despite the economic downturn.

State renewable portfolio standards (RPS) and renewable energy standards (RES) have been significant drivers in the growth of investment in renewable generation. An RPS requires a certain percentage of energy sales (MWh) to come from renewable resources.

Percentages usually increase incrementally from a base year to an ultimate target. Currently, 29 states plus Washington, D.C., have an RPS and six states have renewable goals without financial penalties for nonachievement. As utilities build more renewable-powered generation, the markets in which they participate continue to address the integration of renewable output into their day-ahead and real-time operations and model expected growth as part of their long-term transmission-planning processes.

To encourage the development of distributed generation (DG), or the production of electricity at the site of consumption, and solar power, 16 states plus Washington, D.C., created RPS carve-outs or set-asides to give an extra boost to these resources, which are not yet cost-competitive with other renewables.

Renewable energy certificates (RECs) allow state regulators to track compliance with mandatory RPS targets or verify progress in voluntary state renewable programs. They also allow compliance entities to purchase credits – subject to state imposed limits on amount and price - if they have not generated or bought enough renewable energy to meet their annual requirements. Each reported megawatt-hour (MWh) of eligible generation results in a system-issued REC with a unique identification number to prevent double-counting. Each REC includes attributes such as generator location, capacity, fuel-type and source, owner and the date when operations began.

States and local utilities offer a variety of financial incentives for renewable energy to complement policy mandates. These include tax credits for in-state manufacture of renewable energy equipment, consumer rebates for purchase and installation of renewable generation or production incentives. Production incentives include extra credits for solar output based on RPS solar set-asides and feed-in tariffs.

Five states, two municipal utilities and Ontario, Canada, adopted or expanded feed-in tariffs (FITs) in 2009 to support their energy and environmental goals. Also called feed-in rates or advanced renewable incentives, these programs typically are designed to encourage development of new small- and medium-sized renewable generation projects by residential and independent commercial developers.

FITs require utilities to buy the renewable generation at a fixed rate that is higher than that provided to other generators, under multiyear contracts. This enables smaller distributed renewable generators to avoid having to participate in renewable portfolio standard (RPS) auctions or other competitive procurements and compensates them for more expensive technologies. The utility passes the costs of the program to its customers.

Transmission

The alternating current (AC) power grid operates like an interconnected web, where, with a few exceptions, the flow of power is not specifically controlled by the operators on a line-by-line basis. Instead, power flows from sources of generation to consumers across any number of lines simultaneously, following the path of least resistance. There are a limited number of direct current (DC) lines, which are set up as specific paths with definite beginning and end points for scheduling and moving power. These lines are controllable by operators and have other characteristics that make them attractive to grid planners and operators, such as providing greater grid stability and lower line losses. However, DC lines cost significantly more than AC lines to construct. Consequently, DC lines are typically built for certain specialized applications involving moving large amounts of power over long distances, such as the Pacific Intertie, which extends between the Northwest and California.

Transmission lines provide a certain amount of resistance to the flow of power as electricity travels through them. This resistance is not unlike the wind resistance that a car must overcome as it travels along a highway. The resistance in power lines creates losses: the amount of power injected into a power line diminishes as it travels through the line. The amount of these losses is contingent on many factors, but typically equals several percent of the amount put into the system.

Transmission Service

FERC requires that public utilities that own transmission lines used in interstate commerce offer transmission service on a nondiscriminatory basis to all eligible customers. The rates and terms of service are published in each utility's Open Access Transmission Tariff (OATT). One type of service is point-to-point service. This service involves paying for and reserving a fixed quantity of transmission capacity and moving power up to the reservation amount from one location, the point of receipt (POR), to another location, the point of delivery (POD). Depending on availability, customers may purchase point-to-point service for durations of one hour to multiple years. The price for the service is cost-based and published in the OATT. In cases where there are multiple parties desiring transmission, it is allocated to the party willing to purchase it for the longest period of time. Capacity reassignment is the term for the resale of point-to-point transmission capacity in the secondary market.

Transmission holders may want to sell capacity in the secondary market because it is unneeded, or to make a profit. Capacity reassignment has been permitted since 1996. Beginning in 2007, resellers have been permitted to charge market-based prices for capacity reassignments, as opposed to the original cost-based price at which they purchased the capacity. The number of capacity reassignments increased from around 200 in 2007 to almost 32,000 in 2009. Most of the transactions were hourly, although capacity can also be reassigned on a daily, monthly or yearly basis.

If the market price of energy is greater at the POD than at the POR, the transmission has value. The transmission holder can capture this value by using the transmission – buying energy at the POR, moving it to the POD and selling it. Alternatively, the transmission holder can sell the transmission through a capacity reassignment. Thus, the price of a capacity reassignment should be equal to the expected price differential between the POD and the POR.

Grid Operations

Grid operators dispatch their systems using the least costly generation consistent with the constraints of the transmission system and reliability requirements. The dispatch process occurs in two stages: day-ahead unit commitment, or planning for the next day's dispatch, and economic dispatch, or dispatching the system in real time.

Day-Ahead Unit Commitment

In the unit commitment stage, operators decide which generating units should be committed to be online for each hour, typically for the next 24-hour period. This is done in advance of real-time operations because some generating units require several hours lead time before they are brought online. In selecting the most economic generators to commit, operators take into account forecast load requirements and each unit's physical operating characteristics, such as how quickly output can be changed, maximum and minimum output levels and the minimum time a generator must run once it is started. Operators must also take into account generating unit cost factors, such as fuel and nonfuel operating costs and the cost of environmental compliance.

Also, forecast conditions that can affect the transmission grid must be taken into account to ensure that the optimal dispatch can meet load reliably. This is the security aspect of commitment analysis. Factors that can affect grid capabilities include generation and transmission facility outages, line capacities as affected by loading levels and flow direction and weather conditions. If the security analysis indicates that the optimal economic dispatch cannot be carried out reliably, relatively expensive generators may have to replace less-expensive units.

System and Unit Dispatch

In the system dispatch stage, operators must decide in real time the level at which each available resource from the unit commitment stage should be operated, given the actual load and grid conditions, so that overall production costs are minimized. Actual conditions will vary from those forecast in the day-ahead commitment, and operators must adjust the dispatch accordingly. As part of real-time operations, demand, generation and interchange (imports and exports) must be kept in balance to maintain a system frequency of 60 hertz. This is typically done by automatic generation control (AGC) to change the generation dispatch as needed.

The chart below is a depiction of the supply curve of the power plants for the New York ISO (NYISO). This is also commonly called the supply stack. In it, all of the plants in the NYISO system are shown sorted according to their marginal cost of production. Their cost of production is shown on the vertical axis. The cheapest ones to run are to the left and the most expensive to the right.

Dispatch in the NYISO, for example, first calls on wind plants, followed successively by hydro, nuclear and coal-, gas- and oil-fired generators. This assumes that the plants have sufficient resources – enough wind for the wind powered generators or enough river flow for the hydroelectric plants, for example – and that sufficient transmission capability exists to deliver plant output and meet reliability needs.

In addition, transmission flows must be monitored to ensure that flows stay within voltage and reliability limits.

If transmission flows exceed accepted limits, the operator must take corrective action, which could involve curtailing schedules, changing the dispatch or shedding load. Operators may check conditions and issue adjusted dispatch instructions as often as every five minutes.

Ancillary Services

Ancillary services maintain electric reliability and support the transmission of electricity. These services are produced and consumed in real-time, or in the very near term. NERC and regional entities establish the minimum amount of each ancillary service that is required for maintaining grid reliability.

Regulation matches generation with very short-term changes in load by moving the output of selected resources up and down via an automatic control signal, typically every few seconds. The changes are designed to maintain system frequency at 60 hertz. Failure to maintain a 60- hertz frequency can result in collapse of an electric grid.

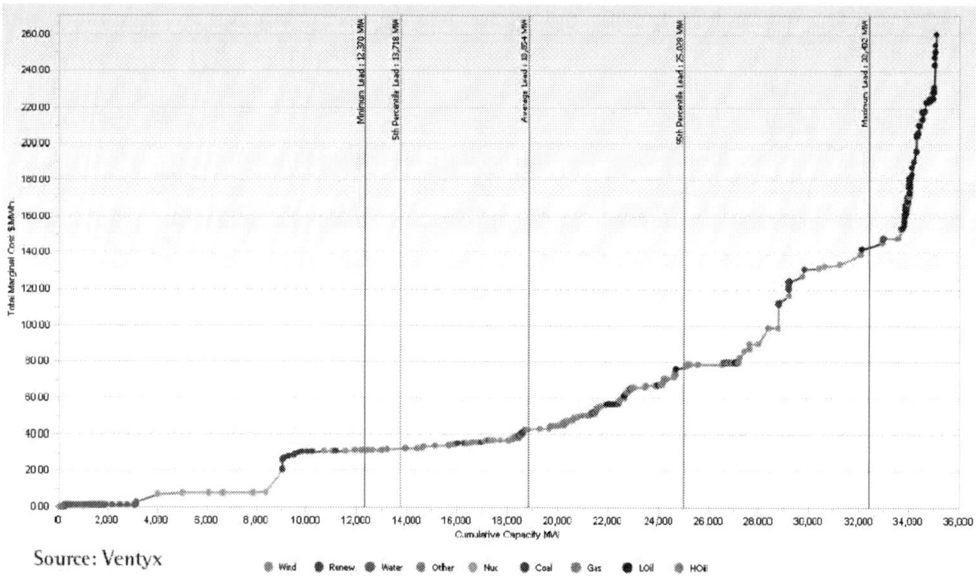

Source: Ventyx

New York ISO Supply Stack.

Operating reserves are needed to restore load and generation balance when a generating unit trips off line. Operating reserves are provided by generating units and demand resources that can act quickly, by increasing output or reducing demand, to make up a generation deficiency. There are three types:

1) ***Spinning reserves*** are primary. To provide spinning reserve a generator must be on line (synchronized to the system frequency) with some unloaded (spare) capacity and be capable of increasing its electricity output within 10 minutes. During normal operation these reserves are provided by increasing output on electrically synchronized equipment or by reducing load on pumped storage hydroelectric facilities. Synchronized reserve can also be provided by demand-side resources.
2) ***Nonspinning reserves*** come from generating units that can be brought online in 10 minutes. Nonspinning reserve can also be provided by demand-side resources.
3) ***Supplemental reserves*** come from generating units that can be made available in 30 minutes and are not necessarily synchronized with the system frequency. Supplemental reserves are usually scheduled in the day-ahead market, allowing generators to offer their reserve energy at a price, thus compensating cleared supply at a single market clearing price. This only applies to ISO/RTOs, and not all reliability regions have a supplemental reserve requirement.

Black start generating units have the ability to go from a shutdown condition to an operating condition and start delivering power without any outside assistance from the electric grid. Hydroelectric facilities and diesel generators have this capability. These are the first facilities to be started up in the event of a system collapse or blackout to restore the rest of the grid.

Reactive power. Electricity consists of current, the flow of electrons, and voltage, the force that pushes the current through the wire. Reactive power is the portion of power that establishes and maintains electric and magnetic fields in AC equipment. It is necessary for transporting AC power over transmission lines, and for operating magnetic equipment, including rotating machinery and transformers. It is consumed by current as it flows. As the amount of electricity flowing on a line increases, so does the amount of reactive power needed to maintain voltage and move current. Power plants can produce both real and reactive power, and can be adjusted to change the output of both. Special equipment installed on the transmission grid is also capable of injecting reactive power to maintain voltage.

Weather

Weather is the single greatest driver of electric demand and, thus, is a major factor in grid operations. System operators therefore rely heavily on weather forecasts to ensure they have the right generation in the right locations to run the grid reliably. Weather affects grid operations in other ways, as well. Primary among these is on the productivity of certain types of power generators: wind and hydroelectric. Wind turbines' power output changes with wind availability and speed, which affects cost of wholesale power. Hydroelectric plants rely on rain and snowfall to provide the river flow needed for their output. Geographically, this is most important in the Pacific Northwest, where seasonal hydro plant output is a critical source of power. Rain and the melting of winter snowpack feed the Columbia and Snake river systems. Surplus power from these generators is typically exported to California to help meet summer peak demand and provide a combination of increased reliability and lower prices. Temperature can also affect the output of other power plants and capacity of transmission lines. Specifically, thermal plants that use a turbine – coal, gas, oil and nuclear plants – become less efficient at higher temperatures. Additionally, the capacity of transmission lines is limited by heat because the conductive material used in fabrication becomes more electrically resistant as they heat up, limiting their throughput.

Markets and Trading

Overview

Markets for delivering power to consumers in the United States are split into two systems: traditional regulated markets and market-regulated markets run by regional transmission organizations (RTOs), which include independent system operators (ISOs).

In general, RTOs use their markets to make operational decisions, such as generator dispatch. Traditional systems rely on management to make those decisions, usually based on the cost of using the various generation options.

Trading for power is also split into over-the-counter (OTC) or bilateral transactions, and RTO transactions. Bilateral transactions occur in both traditional systems and in RTO regions, but in different ways.

Pricing in both RTO and traditional regions incorporate both cost-of-service and market-based rates.

Bilateral Transactions

Bilateral or OTC transactions between two parties do not occur through an RTO. In bilateral transactions, buyers and sellers know the identity of the party with whom they are doing business.

Bilateral deals can occur through direct contact and negotiation, through a voice broker or through an electronic brokerage platform, such as the IntercontinentalExchange (ICE). The deals can range from standardized contract packages, such as those traded on ICE, to customized, complex contracts known as structured transactions.

Whether the trade is done on ICE, directly between parties or through another type of broker, the trading of standard physical and financial products, such as next-day on-peak firm or swaps, allows index providers to survey traders and publish price indexes. These indexes provide price transparency.

Physical bilateral trades involving the movement of the energy from one point to another require that the parties reserve transmission capacity to move the power over the transmission grid. Transmitting utilities are required to post the availability of transmission capacity and offer service on an Open Access Same-Time Information System (OASIS) website. Traders usually reserve transmission capacity on OASIS at the same time they arrange the power contract. When it comes time to use the reservation to transfer power between balancing authorities, one of the parties to the transaction submits an eTag electronically to Open Access Technology International (OATI), NERC's eTag contractor.

OATI will process the tag and send it to all parties named on the eTag. This ensures the orderly transfer of energy and provides transmission system operators the information they need to institute curtailments as needed. Curtailments may be needed when a change in system conditions reduces the capability of the transmission system to move power and requires some transactions to be cut or reduced. Bilateral physical transactions conducted in RTOs are settled financially. Generators offer their power into the RTO markets, and load is served through the power dispatched by the RTO. The RTO then settles bilateral transactions based on the prices in the contracts and the prices that occurred in the RTO markets.

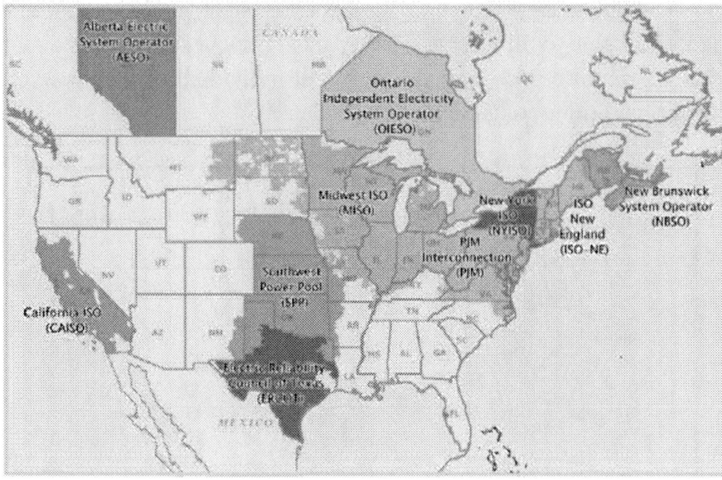

Source: Energy Velocity.

North American Regional Transmission Organizations.

Cost-Based Rates

Cost-based rates are used to price most transmission services and some electricity when the Commission determines that market-based rates are not appropriate, or when an entity does not seek market-based rate authority. Cost-based rates are set to recover costs associated with providing service and give a fair return on capital. Cost-based rates are typically listed in a published tariff.

The following are major inputs to setting cost-based electricity rates:

- Determining used-and-useful electricity plants. This may include generation facilities, transmission facilities, distribution plants and office and related administration facilities.
- Determining expenses from the production, transmission and distribution of electricity, including fuel and purchased power, taxes and administrative expenses.
- Establishing a fair return on capital, known as the cost of capital. This includes determining the cost of debt, common equity, preferred stock and commercial paper and other forms of short-term borrowing such as lines of credit used to finance projects and provide cash for day-to-day operations.
- Allocating electric plant and other expenses among various customer classes and setting the rate structure and rate levels.

Market-Based Rates

Under market-based rates, the terms of an electric transaction are negotiated by the sellers and buyers in bilateral markets or through RTO market operations. The Commission grants market-based rate authority to electricity sellers that demonstrate that they and their affiliates lack or have adequately mitigated horizontal market power (percent of generation owned relative to total generation available in a market), and vertical market power (the ability to influence the cost of production for competitive electricity suppliers). Wholesale sellers who have market-based rate authority and who sell into day-ahead or real-time markets administered by a RTO do so subject to the specific RTO market rules approved by the Commission and applicable to all market participants. Thus, a seller in such markets not only must have an authorization based on analysis of that individual seller's market power, but it must abide by additional rules contained in the RTO tariff.

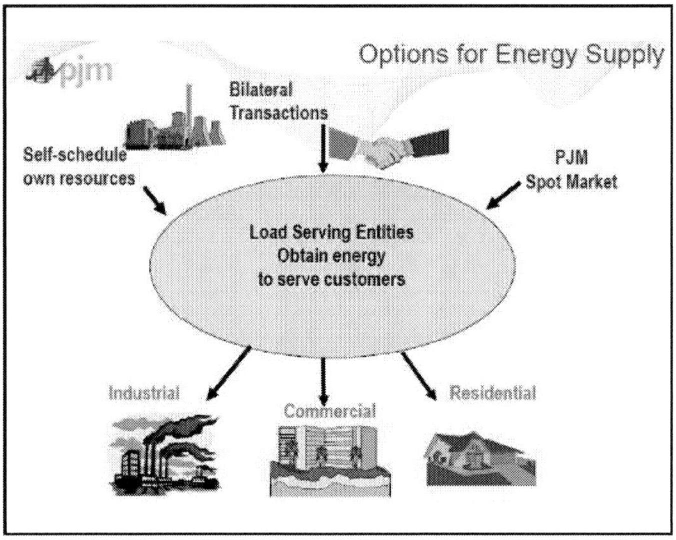

Supplying Load

Suppliers serve customer load through a combination of self-supply, bilateral market purchases and spot purchases. In addition to serving load themselves, load-serving entities (LSEs) can contract with others to do so. The choices are:

- Self-supply means that the supplying company generates power from plants it owns to meet demand.
- Supply from bilateral purchases means that the load-serving entity buys power from a supplier.
- Supply from spot RTO market purchases means the supplying company purchases power from the RTO.

LSEs' sources of energy vary considerably. In ISO-NE, NY-ISO and CAISO, the load-serving entities divested much or all of their generation. In these circumstances, LSEs supply their customers' requirements through bilateral and RTO market purchases. In PJM, MISO and SPP, load-serving entities may own significant amounts of generation either directly or through affiliates and therefore use self-supply as well as bilateral and RTO market purchases.

Traditional Power Markets

Traditional wholesale electric markets exist primarily in the Southeast, Southwest and Northwest. About 40 percent of all retail customers are in traditional wholesale markets where utilities are responsible for system operations and management, and, typically, for providing power to retail consumers. Utilities in these markets are frequently vertically integrated – they own the generation, transmission and distribution systems used to serve electric consumers.

They may also include federal systems, such as the Bonneville Power Administration, the Tennessee Valley Authority and the Western Area Power Administration. Utilities in traditional regions have the following responsibilities:

- Generating or obtaining the power needed to serve customers (this varies by state);
- ensuring the reliability of its transmission grid;
- balancing supply and demand instantaneously;
- dispatching its system resources as economically as possible;
- coordinating system dispatch with neighboring balancing authorities;
- planning for transmission requirements within the utility's footprint; and
- coordinating its system development with neighboring systems.

Wholesale physical power transactions occur through bilateral markets.

Regional Markets

Introduction

Two-thirds of the population of the United States and more than one-half of Canada's population are served by electricity markets run by regional transmission organizations or independent system operators (RTOs/ISOs). There is little practical distinction between a RTO and an ISO.

The main distinction between RTO/ISO markets and their predecessors (such as vertically integrated utilities, municipal utilities and co-ops) is that RTO/ISO markets deliver reliable electricity through competitive market mechanisms.

The basic functions of a RTO or ISO include the following:

- Ensure the reliability of the transmission grid;
- operate the grid in a defined geographic footprint;
- balance supply and demand instantaneously;
- operate competitive nondiscriminatory electricity markets;
- provide nondiscriminatory interconnection service to generators; and
- plan for transmission expansion on a regional basis.

In performing these functions, RTOs have operational control of the transmission system, are independent of their members, coordinate the maintenance of generation and transmission system, and oversee a transmission planning process to identify needed upgrades in both the near- and long-term.

RTOs/ISOs do not own transmission or generation assets; perform the actual maintenance on generation or transmission equipment; or directly serve end use customers.

Currently, seven RTOs/ISOs operate in the United States, listed below in order of the size of their peak load:

- PJM Interconnection (PJM); 145 GW (summer)
- Midwest ISO (MISO); 137 GW (summer)
- Electric Reliability Council of Texas (ERCOT); 63 GW (summer)
- California ISO (CAISO); 50 GW (summer)
- Southwest Power Pool (SPP); 47 GW (summer)
- New York ISO (NYISO); 34 GW (summer)
- New England ISO (ISO-NE); 27 GW (summer)

Market Operations

RTO operations encompass multiple services that are needed to provide reliable and economically efficient electric service to customers. Each of these services has its own parameters and pricing. The RTOs use markets to determine the provider(s) and prices for many of these services. These markets include the day-ahead energy market (sometimes called a Day 2 market), real-time energy market (sometimes called a Day 1 or balancing market), capacity markets (designed to ensure enough generation is available to reliably meet peak power demands), ancillary services markets, financial transmission rights (contracts for hedging the cost of limited transmission capability) and virtual trading (financial instruments to create price convergence in the day-ahead and real-time markets).

RIO Energy Markets

All but one RTO (i.e., SPP) electricity market has day-ahead and real-time markets. The day-ahead market schedules electricity production and consumption before the operating day, whereas the real-time market (also called the balancing market) reconciles any differences between the schedule in the day-ahead market and the real-time load while observing reliability criteria, forced or unplanned outages and the electricity flow limits on transmission lines.

The day-ahead energy market produces financially binding schedules for the production and consumption of electricity one day before its production and use (the operating day). The purpose of the day-ahead market is to give generators and load-serving entities a means for scheduling their activities sufficiently prior to their operations, based on a forecast of their needs and consistent with their business strategies.

In day-ahead markets, the schedules for supply and usage of energy are compiled hours ahead of the beginning of the operating day. The RTO then runs a computerized market model that matches buyers and sellers throughout the geographic market footprint for each hour throughout the day. The model then evaluates the bids and offers of the participants, based on the power flows needed to move the electricity throughout the grid from generators to consumers. Additionally, the model must account for changing system capabilities that occur based on weather and equipment outages, plus rules and procedures that are used to ensure system reliability. The market rules dictate that generators submit supply offers and loads submit demand bids to the RTO by a deadline that is typically in the morning of the day-ahead scheduling. Typically, 95 percent of all energy transactions are scheduled in the day-ahead market, and the rest scheduled in real-time.

Generation and demand bids that are scheduled by the day-ahead market are settled at the day-ahead market prices. Inputs into setting a day-ahead market schedule include:

- Generator offers to sell electricity each hour;
- bids to buy electricity for each hour submitted by load-serving utilities;
- demand-response offers by customers to curtail usage of electricity;
- virtual demand and supply offers; and
- operational information about the transmission grid and generating resources, including planned or known transmission and generator outage, the physical characteristics of generating resources including minimum and maximum output levels and minimum run time and the status of interconnections to external markets.

The real-time market is used to balance the differences between the day-ahead scheduled amounts of electricity based on day-ahead forecast and the actual real-time load. The real-time market is run hourly and in 5-minute intervals and clears a much smaller volume of energy and ancillary services than the day-ahead market, typically accounting for only 5 percent of scheduled energy. For generators, the real-time market provides additional opportunities for offering energy into the market. Megawatts over- or under-produced relative to the day-ahead commitments are settled at real-time prices.

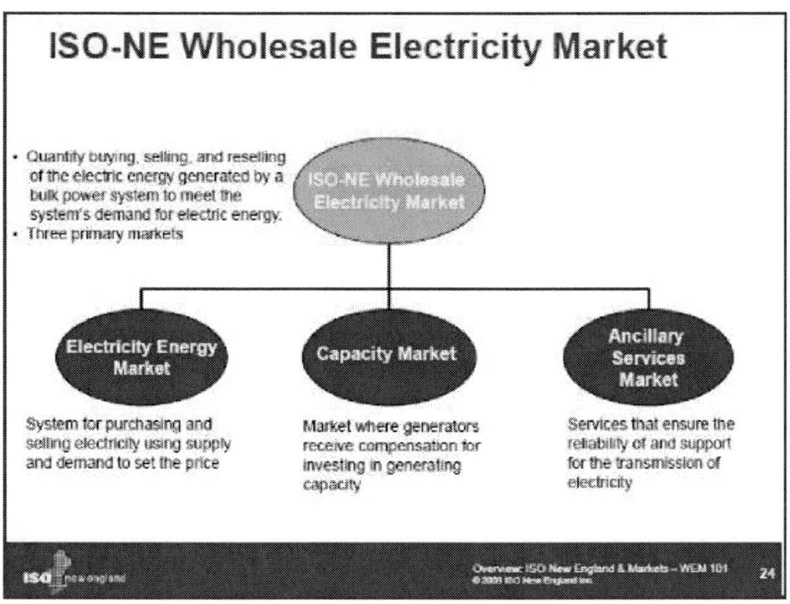

Real-time market prices are significantly more volatile than the day-ahead market prices. This stems from demand uncertainty, transmission and generator forced outages and other unforeseen events. Because the day-ahead market generally is not presented with these events, it produces more stable prices than in real-time. Also, because the volumes in the real-time market are much smaller, there is an increased likelihood of supply and demand imbalances, which lead to both positive and negative price movements.

RTOs use markets to deal with transmission constraints through locational marginal pricing (LMP).

The RTO markets calculate a LMP at each location on the power grid. The LMP reflects the marginal cost of serving load at the specific location, given the set of generators that are being dispatched and the limitations of the transmission system. LMP has three elements: an energy charge, a congestion charge and a charge for transmission system energy losses.

If there are no transmission constraints, or congestion, LMPs will not vary significantly across the RTO footprint. Transmission congestion occurs when there is not enough transmission capacity for all of the least-cost generators to be selected. The result is that some more expensive generation must be dispatched to meet demand, units that might not otherwise run if more transmission capacity were available.

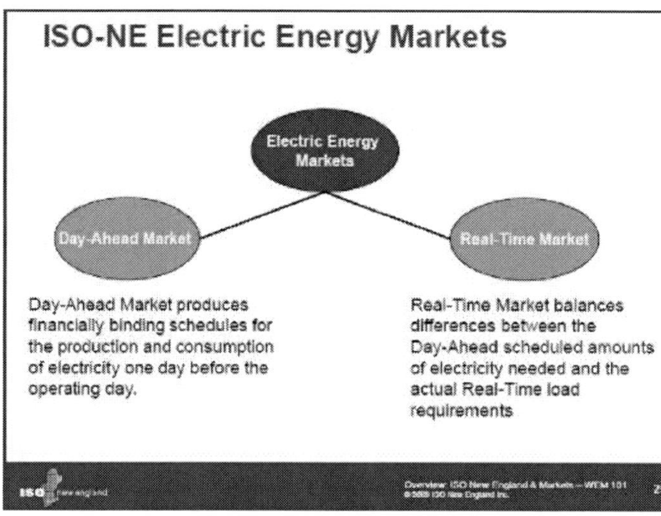

When there are transmission constraints, the highest variable cost unit that must be dispatched to meet load within transmission-constrained boundaries will set the LMP in that area. All sellers receive the LMP for their location and all buyers pay the market clearing price for their location.

The primary means used for relieving transmission congestion constraints is by changing the output of generation at different locations on the grid. The market-based LMP sends price signals that reflect congestions costs to market participants. That is, LMPs take into account both the impact of specific generators on the constrained facility and the cost to change (redispatch) the generation output to serve load. This change in dispatch to is known as security constrained redispatch (see chart).

This redispatch could be implemented by using nonmarket procedures such as transmission loading relief (TLR). North American Electric Reliability Corp. (NERC) established the TLR process for dealing with reliability concerns when the transmission network becomes overloaded and power flows must be reduced to protect the network. A TLR is used to ration transmission capacity when the demand for transmission is greater than the available transmission capacity (ATC). The rationing is a priority system that cuts power flows based on size, contractual terms and scheduling.

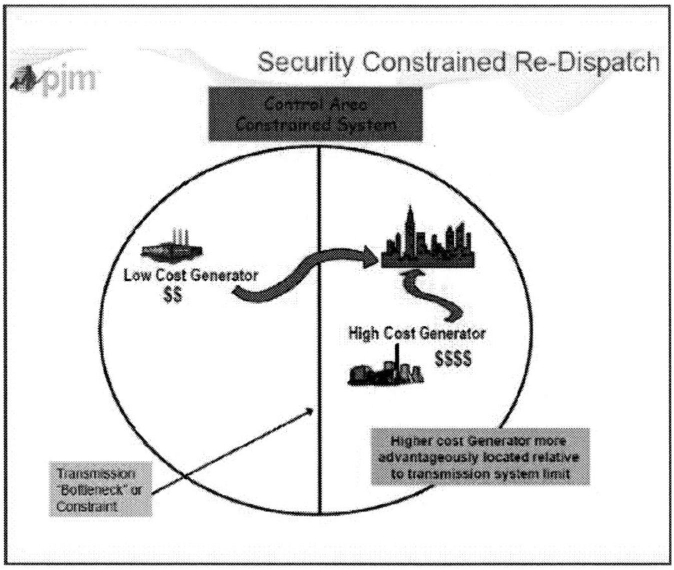

Scarcity pricing is a mechanism used by RTOs and ISOs to send price signals in the real-time market when there is a systemwide shortage of power reserves. These events occur when there is a shortage of power to meet system requirements to meet load and provide sufficient backup reserves. This can be caused by unexpectedly high power loads, supply disruptions or both.

RTOs follow one of four approaches to ensure that the market price for energy accurately reflects the value of energy during shortage periods:

- Increase the allowed bidding price of energy supply above normal levels during an emergency;
- Increase bid caps above the current level during an emergency for demand bids, while keeping generation offer caps in place;
- Establish a pricing structure for operating reserves that would raise prices as operating reserves grow short (demand curve); and
- Set the market-clearing price during an emergency for all supply and demand response resources dispatched equal to the payment made to participants in an emergency demand-response program.

Reliability must-run (RMR) units are generating plants that would otherwise retire but the RTO has determined they are needed to ensure reliability. They could also be units that have market power due to their location on the grid. RTO/ ISOs enter into cost-based contracts with these generating units and allocate the cost of the contract to transmission customers. In return for payment, the RTO may call on the owner of an RMR generating unit to run the unit for grid reliability. The payment must be sufficient to pay for the cost of owning and maintaining the unit even if it does not operate.

The reason for developing capacity markets (described below) is in part to compensate generation owners for keeping these units in service where necessary, in addition to prompting the construction of new generation and use of demand response by consumers. Transmission upgrades can also reduce the need for RMR units by increasing generation deliverability throughout the RTO.

RTO Capacity Markets

RTOs, like other electric systems, are required to maintain adequate generation reserves to ensure that sufficient generation and demand-resource capacity are available to meet load and reliability requirements. LSEs have typically satisfied their reserve obligations with owned generation or bilateral contracts with other suppliers. Some RTOs have mechanisms to obtain capacity commitments, such as capacity auctions and capacity payments.

Most RTOs run a capacity market to allow LSEs a way to satisfy their reserve obligation. These markets cover short-term capacity, such as a month, season or year. PJM and ISO-NE run capacity auctions up to three years prior to when the capacity is needed. The near-term focus is consistent with providing payments to existing generation, or generation such as combustion turbines that can be sited and built within three years.

Financial Transmission Rights

Financial transmission rights (FTRs) are contracts that give market participants an offset, or hedge, against transmission congestion costs in the day-ahead market. They protect the holder from costs arising from transmission congestion over a specific path on the grid.

FTRs were originally developed in part to give native load-serving entities in the nascent RTOs price certainty similar to that available to traditional vertically integrated utilities operating in non-RTO/ISO markets. This practice continues, as FTRs are allocated to load-serving entities, transmission owners or firm transmission right holders in RTOs based on historical usage, and to entities that fund the construction of specific new transmission facilities. The details of the programs vary by RTO.

FTRs allow customers to protect against the risk of congestion-driven price increases in the day-ahead market in the RTOs and ISOs. Congestion costs occur as the demand for scheduled power over a transmission path exceeds that path's flow capabilities. For example, if the transmission capacity going from Point A (the source) to Point B (the sink) is 500 MW, but the RTO seeks to send 600 MW of power from Point A to Point B when calling on the least-cost generators to serve load, the path will be congested. This will cause the price at the source to decline or the price at sink to increase, or both, causing the congestion cost of serving point B from Point A to increase. By buying an FTR over the path from Point A to Point B, the FTR holder is paid the difference of the congestion prices at the sink and source, thus allowing it to hedge against the congestion costs incurred in the day-ahead market.

FTRs are acquired through allocations and purchases. FTRs can be purchased in the RTO-administered auctions or in the secondary market.

Allocations may stem from a related product, auction revenue right (ARR). ARRs provide the firm transmission capacity holders, transmission owners or LSEs with a portion of the money raised in the FTR auctions. In general, they are allocated based on historic load served and, in some RTOs, can be converted to FTRs. As with FTRs, ARRs, too, give eligible members an offset or hedge against transmission congestion costs in the day-ahead market. If converted to FTRs, the holder gets revenue from congestion. If kept as ARRs, the holder gets revenue from the FTR auction.

The main method for procuring FTRs is through an auction, which typically includes an annual (or multiyear) auction of one-year FTRs and monthly (or semiannually) auctions of shorter-term FTRs provided by existing FTR holders or made available by the RTO. The auctions are scheduled and run by the RTO, which requires bidding parties to post credit to cover the positions taken. FTR auction revenues are used to pay the holders of ARRs and assist the funding of future congestion payments to FTR holders. There is also a secondary market for FTRs (such as PJM's eFTR), but only a small number of transactions have been reported.

The quantity of FTRs made available by the RTO is bounded by the physical limits of the grid, as determined by a simultaneous feasibility test across all potential flowgates. This test is performed by the RTO prior to making FTRs available at auction, and takes into account existing FTR positions and system constraints. The resulting portfolio of FTRs allocated or offered at auction represents an absolute constraint on the size of the net positions that can be held by the market. Participants in FTR auctions can procure counterflow FTRs, which directly offset prevailing flow FTR capacity, thereby allowing the value at risk on a given path to exceed the physical limits of the line. However, such bids are physically constrained, as the net position held on the path must always conform to the simultaneous feasibility test.

Although FTRs are used by transmission providers and load-serving entities as a hedge, they can be purchased by any creditworthy entity seeking their financial attributes either as a hedge or as a speculative investment. In this regard, FTRs are similar to financial swaps that are executed as a contract for differences between two day-ahead LMPs (swaps are explained in the chapter on financial markets). However, FTRs are substantially different from swaps in that the quantity of FTRs is linked to physical constraints in the transmission grid, while the quantity of swaps is not. Further, FTRs are procured by allocation or FTR auction, while swaps are procured through financial over-the-counter markets or exchanges.

Variation in RIO FIRs

Five of the six FERC-jurisdictional RTOs trade FTRs or FTR equivalent products, with SPP planning to use FTRs in its future market design. However, the types and qualities of the rights traded across the organized markets vary, as do differences in the methods used to allocate, auction and transfer these rights. These attributes of the FTR markets are discussed below.

Flow Type: Prevailing Flow and Counterflow. A prevailing flow FTR generally has a source in an historic generation-rich location and a sink that is in a historic load-heavy location. Alternatively, the source of a prevailing flow FTR is on the unconstrained side of a transmission interface and the sink on the constrained side. Auction clearing prices for prevailing flow FTRs are positive. Conversely, a counterflow FTR often has a source in an historic load-heavy location and a sink that in an historic generation-rich location. As a result, auction clearing prices for counterflow FTRs are negative.

Peak Type: On-peak, Off-peak, 24-hour. FTRs can be purchased for either 16-hour on-peak blocks, 8-hour off-peak blocks or around-the-clock. Only PJM offers all three peak type products, whereas ISO-NE, MISO and CAISO offer on-peak and off-peak products. NYISO offers only the 24-hour product.

Allocated Rights. The five RTOs allocate transmission rights to transmission owners or load-serving entities within their markets. In PJM, MISO and ISO-NE, these are allocated as auction revenue rights (ARRs), which give their holders the right to receive a share of the funds raised during the FTR auctions. The CAISO allocates congestion revenue rights (CRR), which provide their holders a stream of payments based on the actual congestion occurring on associated paths. Finally, NYISO allocates both auction-based and congestion-based rights through multiple instruments. PJM and MISO allow ARR holders to convert all of these rights to FTRs; NYISO allows only a portion of ARR-equivalent instruments to be converted to its version of FTRs, called transmission congestion credits (TCCs). ISO-NE does not allow such conversions, while the CAISO's allocation is already in a form equivalent to an FTR. Converted ARRs are fully fungible in PJM, the MISO and NYISO; CAISO only allows the sale of allocated CRRs in its secondary market, and ISO-NE has no converted instruments to sell.

Auctioned Rights. All RTOs provide FTRs (or equivalent CRRs or TCCs) for sale to the public through two or more auctions held at various times of the year. The products sold vary by market and by auction, with some products made available only at specific auctions.

Secondary Markets. With the exception of the NYISO, each of the markets that auction FTRs also operates a bulletin board or similar venue designed to enable a secondary trading platform for FTRs. However, none of these platforms has had significant volume. NYISO offered to create a bulletin board for its participants if requested, but received no requests. The CAISO is the only market that requires the reporting of secondary FTR transactions; such transactions have not occurred despite the inability of CRR holders to resell their positions through the auction process.

Virtual Transactions

Virtual bids and offers (collectively, virtuals) are used by traders participating in the RTO markets to profit from differences between day-ahead and real-time prices. The quantity of megawatts (MW) purchased or sold by the trader in the day-ahead market is exactly offset by a sale or purchase of an identical quantity of MW in the real-time, so that the net effect on the market quantity traded is zero.

Virtual trading is allowed in the five RTOs with day-ahead and real-time markets.

Although a trader does not have to deliver power, the transaction is not strictly financial. Virtual transaction can physically set the LMPs, the basis for payments to generators or from load.

For each hour, net virtual trades are added to the demand forecast for load if virtual demand is greater than virtual supply. This has the effect of raising the price in the day-ahead market and, more importantly, increasing the amount of generation resources procured by the RTO/ISO. Since these resources will be available to the real-time market, the failure of the virtual load to materialize will decrease the real-time demand below forecast, thus placing downward pressure on real-time prices. The placement of virtuals affects the dispatch of physical capacity.

The primary benefits of virtual transactions are achieved through their financial impact on the markets. Virtuals sometimes are referred to as convergence bidding, as a competitive virtual market should consistently cause the day-ahead and real time prices to converge in each hour.

The convergence of day-ahead and real-time prices within the RTOs is is intended to mitigate market power and improve the efficiency of serving load. Thus, virtuals have a physical impact upon the operations of the RTO, as well as on market participants that physically transact at the LMPs set in the day-ahead and real-time markets.

Transmission Operations

Each RTO's Open Access Transmission Tariff (OATT) specifies the transmission services that are available to eligible customers. Customers submit requests for transmission service through the Open Access Same-Time Information System (OASIS). RTOs evaluate each transmission-service request using a model of the grid called a state estimator. Based on the model's estimation of the effects on the system, the request for transmission service is either approved or denied.

Transmission operators, including RTOs, offer two major types of transmission service: point-to-point service and network service. Network service generally has priority over point-to-point service. RTOs work with transmission owners to plan and coordinate the operation, maintenance and expansion of transmission facilities in order to provide network and point-to-point customers with transmission service.

Transmission Rights by Transmission Grid

	PJM	**MISO**	**ISO-NE**	**NYISO**	**CAISO**
Name for Allocated Transmission Rights	***ARR*s**	***ARR*s**	***ARR*s**	Multiple	CRR
Convertible to Congestion Rights?	Yes	Yes	No	Some	-
Name for Auctioned Congestion Rights	***FTR*s**	***FTR*s**	***FTR***	*TCCs*	*CRRs*
Congestion Right Auction Format					
Annual	Yes	Yes	Yes	No	Yes
Semiannual	No	No	No	Yes	No
Monthly	Yes	Yes	Yes	Yes	Yes
Congestion Right Auction Products:					
Multiyear	Yes	No	No	No	No
Annual	Yes	No	Yes	Yes	No
Semiannual	No	No	No	Yes	No
Quarterly	Yes	Yes	No	No	Yes
Monthly	Yes	Yes	Yes	Yes	Yes
Auction Allows Participant Resale?	Yes	Yes	Yes	Yes	No
Congestion Right Options	Yes	No	No	No	No
Formal Secondary ***FTR*** Market?	Yes	Yes	Yes	No	Yes

Network transmission service is used for the transmission of energy from network generating resources to an RTO's network loads.

- Network transmission service enables network customers to use their generation resources to serve their network loads in a RTO.
- Network customers also can use the service to deliver economy energy purchases to their network loads.

Point-to-point transmission service uses an RTO's system for the transmission of energy between a point of receipt and a point of delivery, which can be into, out of, or through the RTO's Control Area. RTOs offer firm and non-firm point-to-point transmission service for various lengths of time.

- Firm service has reservation priority over nonfirm point-to-point service.
- Nonfirm point-to-point transmission service is provided from the available transmission capability beyond network and firm point-to-point transmission service.

Transmission Planning

RTOs have systemwide or regional planning processes that identify transmission system additions and improvements needed to keep electricity flowing. Studies are conducted that test the transmission system against mandatory national reliability standards as well as regional reliability standards. The North American Electric Reliability Corp. (NERC) is the organization responsible for setting national reliability standards.

RTO transmission planning studies may look 10-15 years into the future to identify transmission overloads, voltage limitations and other reliability problems. RTOs then develop transmission plans in collaboration with transmission owners to resolve potential problems that could otherwise lead to overloads and blackouts. This process culminates in one recommended plan for the entire RTO footprint.

Financial Policies

Financial settlement is the process through which payments due from customers and to generators are calculated. Market settlements depend on day-ahead schedules, real-time metering, interchange schedules, internal energy schedules, ancillary service obligations, transmission reservations, energy prices, FTR positions and capacity positions. For each market participant a customer invoice of charges and credits includes the costs of services used to serve load.

Generally, customers receive weekly or monthly invoices stating their charges and credits. Weekly invoices must be settled within a few days of being issued, while monthly invoices must be paid within either one or two weeks depending on the policies of each RTO. All payments are made electronically. Disbursements are made within several days of the date payments are due.

Credit Policies

Defaults by market participants in RTOs have generally been socialized, meaning that the cost is spread across the market. To minimize this risk, RTOs have credit policies in their tariffs, which contain provisions related to credit evaluations, credit limits, forms of collateral and the consequences of violations or defaults.

Regions

Markets vary around the United States by market type – traditional or RTO – generation types, customer use, climate, fuel costs, political and regulatory conditions, and other factors. Consequently, prices vary, driven by these market factors.

Southeast Wholesale Market

The Southeast electric market is a bilateral market that includes all or parts of Florida, Georgia, Alabama, Mississippi, Louisiana, Arkansas, Tennessee, North Carolina, South Carolina, Texas, Missouri and Tennessee. It encompasses all or part of two NERC regions: the Florida Reliability Coordinating Council (FRCC) and the Southeastern Electric Reliability Council (SERC). Major hubs include Entergy, Southern and TVA.

Southeastern power markets have their roots in the 1960s. In the wake of the Northeast Blackout of 1967, the Southeast began to build out its electric transmission grid; there now are several large transmission lines connecting large power plants to the grid. This was primarily to ensure reliability, but it also had economic consequences. Increased integration allowed utilities to more effectively share reserves, as well as the costs and risks of new plant construction.

If a utility were building a large nuclear or coal- fired generating facility, it would be cost-effective to have reserve sharing agreements with neighboring systems that provided the backup or capacity reserves, rather than building reserves individually. In addition, a stronger grid allowed the output of large power plants to be deliverable throughout the region, thus allowing more than one utility to share in the ownership and the costs of building large new plants. This reduced the financial risks associated with ownership of large new generating facilities to any single utility, thus making ownership of large base-load coal and nuclear units more affordable to the utilities and less risky.

A stronger transmission system also allowed for more economic transactions, including both spot transactions and long-term firm power deliveries. External sales resulted in more efficient use of grid resources and reduced costs to both buyers and sellers.

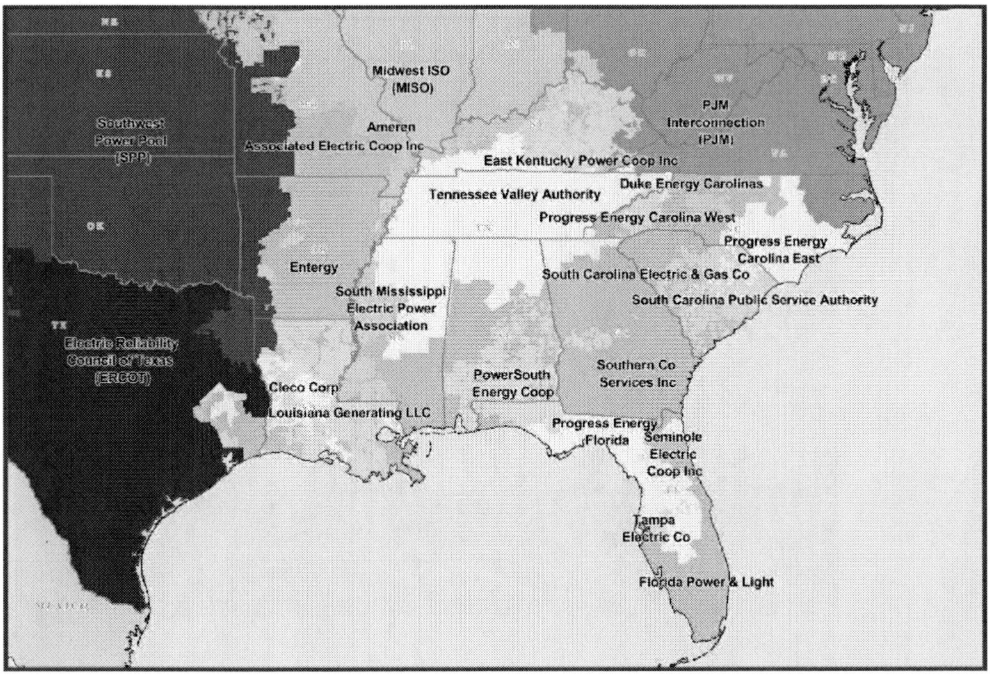

Southeast Electric Regions.

Resource Base

Within the Southeast, the resource mix varies between the two NERC subregions. The FRCC uses more gas- and oil- fired generation than the rest of the Southeast, and it is the only area where oil is significantly employed. Gas is the marginal fuel in almost all hours in the FRCC. Within SERC, the Southern subregion has historically generated as much as 85 percent of its electricity from baseload coal and nuclear plants. In recent years, natural gas used for generating electricity has become increasingly popular. The pattern began to change as gas supplies increased and prices fell and natural gas-fired power plants began to displace older, less-efficient coal-fired generation.

The Entergy subregion uses gas to a much greater extent than the regional average; it is the marginal fuel more than 70 percent of the time. The TVA subregion has a significant amount of hydro and nuclear capacity and output, and very little dependence on gas. The VACAR subregion has the highest utilization of nuclear generation in the Southeast; 94 percent of this subregion's output is from baseload coal and nuclear facilities.

Trading and Markets

Physical and financial electricity products are traded using Entergy, Southern, TVA, VACAR and Florida price points. Volumes for these products remain low, especially in Florida, where merchant power plant development is restricted by a state statute.

Virtually all the physical sales in the Southeast are done bilaterally. Long-term energy transactions appear to be a hallmark of the Southeast; wholesale electricity transactions for a year or more outweigh spot transactions. Many long-term agreements involve full-requirements contracts or long-term purchase power agreements. Spot transactions accounted for less than one percent of overall supply and tend to occur during periods of system stress, usually summer heat waves or winter cold snaps. Even for a large company such as Southern Co., spot transactions occurred less than 30 percent of the time.

Wholesale spot power markets in the Southeast have little spot trading and lack transparency. The relative lack of spot trades yields little data on which to base price reporting. ICE reports no electric power price for Florida. And while another publisher reports one spot electric power price for Florida, on most days, there are no reported volumes. Given the bilateral nature of wholesale power transactions in the Southeast, and the small spot market, interest in financial power products in the Southeast is weak. As a result, ICE does not provide a financial swap product in the Southeast.

Despite the bilateral nature of the wholesale trade and the small size of the spot market, marketers do have some presence in the Southeast. For example, Constellation Energy Commodities Group contributes to the trading in the Southern region, being a participant in 41 percent of total marketer related sales. While Constellation does not own generation in Southern, it does have several multiyear agreements with generating units.

Unique Market Features

Southern Co. Auction

Since April 23, 2009, Southern Co. has been holding daily and hourly auctions for power within its balancing area. This balancing area encompasses the service territories of Southern Co. utilities: Georgia Power, Alabama Power, Mississippi Power and Gulf Power.

According to the auction rules, Southern must offer all of its available excess generation capacity into the auction, after regulation and contingency reserves are met. The offer prices are capped because the auction is intended to mitigate any potential ability of Southern to withhold its generation resources within its balancing area.

The products auctioned are day-ahead power and real-time power (an auction takes place an hour ahead of when the energy is scheduled to flow).

Offers to sell energy and bids to purchase energy are evaluated using the simple method of sorting offers in ascending order and bids in descending order.

The auction matches parties to facilitate a bilateral transaction that is ultimately independent of the auction. Thus, there is no collateral requirement necessary to participate in the auction. However, credit screening rules dictate that matches are made only between entities willing and able to do business with one another. The selection process is based on information that each entity submits to the auction administrator.

When the auction began in 2009, Southern Co. was the only participant that could sell into it. On Jan. 3, 2010, other entities were allowed to sell into the auction, and Southern became eligible to make purchases in the auction as well as sales. However, activity in the auction has been sparse since its inception.

Entergy Independent Coordinator of Transmission

Southwest Power Pool (SPP) serves as the independent coordinator of transmission (ICT) for Entergy Services. In this role, SPP oversees the operations of the Entergy transmission system and produces regional planning assessments.

Florida IPP rule

The Florida Public Service Commission's (PSC) competitive bidding rules require investor-owned utilities (IOUs) to issue requests for proposals for any new generating project of 75 MW or greater, exclusive of single-cycle combustion-turbines. The bidding requirement can be waived by the PSC if the IOU can demonstrate that it is not in the best interests of its ratepayers.

Western Regions

The power markets in the western United States are bilateral markets except in most of California. The West includes the Northwest Power Pool (NWPP), the Rocky Mountain Power Area (RMPA) and the Arizona, New Mexico, Southern Nevada Power Area (AZ/NM/SNV) within the Western Electricity Coordinating Council (WECC), a regional entity. These areas contain many balancing authorities (BAs) responsible for dispatching generation, procuring power, operating the transmission grid reliably and maintaining adequate reserves. Although the BAs operate autonomously, some have joint transmission-planning and reserve-sharing agreements.

Physical sales in western states are almost entirely bilateral sales, with a small amount sold into the California ISO's market. Trading in the western states differs from the rest of the country because financial players are active in the physical markets, as well as having a robust financial electricity market.

The volume of financial sales on ICE is roughly as large as physical sales. Physical sales in WECC are dominated by financial and marketing companies.

The NWPP is composed of all or major portions of the states of Washington, Oregon, Idaho, Wyoming, Montana, Nevada and Utah, a small portion of Northern California and the Canadian provinces of British Columbia and Alberta. This vast area covers 1.2 million square miles. It is made up of 20 BAs. The peak demand is 54.5 GW in summer and 63 GW in winter. There is 80 GW of generation capacity, including 43 GW of hydroelectric generation.

Resources

The NWPP has a unique resource mix. Hydro generation is more than 50 percent of power supply, compared to the U.S. average of only 6 percent of power supply. The hydro generation is centered around many dams, mostly on the Columbia River. The largest dam, the Grand Coulee, can produce as much power as six nuclear plants. Due to the large amount of hydroelectric generation, the Northwest typically has cheap power and exports power to neighboring regions, especially California, to the extent that there is transmission capacity to carry the power to more expensive markets.

The amount of hydropower produced depends on a number of factors, some natural and some controllable. On a seasonal basis, the intensity and duration of the water flow is driven by snowpack upriver in the mountains, the fullness of the reservoirs and rainfall. On a short-term basis, the power generation is influenced by decisions to release water locally and upstream to generate power, as well as local water-use decisions that have nothing to do with the economics of power generation, but are made for recreation, irrigation and wildlife considerations. The peak generation begins in the spring, when the snow melts, and may last into early summer.

When there is less water available, the Northwest may rely more on its coal and natural gas generation. It will occasionally import power from neighboring regions when loads are high.

Trading and Markets

The water forecast affects the forward market for electricity in the Northwest. The daily water flow as well as weather conditions influence the prices in the daily physical market. When there is an abundance of hydro generation, the Northwest will export as much as possible on the transmission lines leading into California. Sometimes in off-peak hours there is so much generation that power prices are negative because the transmission lines are full and there is not enough local load to take all of the power.

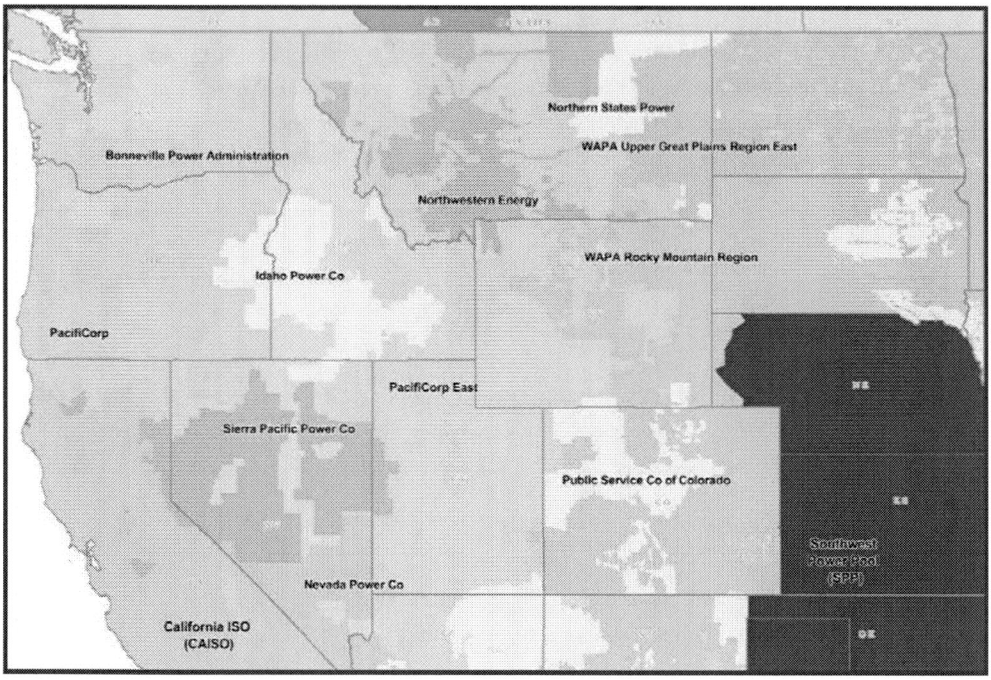

Northwest Electric Market.

The largest seller of wholesale power is the Bonneville Power Administration (BPA), a federal agency that markets the output from federally owned hydroelectric facilities and owns 75% of the region's high-voltage transmission. It meets approximately one-third of the region's firm energy supply, mostly with power sold at cost. BPA gives preference to municipal and other publicly owned electric systems in allocating its output.

Both the Alberta IESO and British Columbia Hydro are members of the NWPP. Net interchange between these two BAs and the United States tends to result in net exports from the United States into Canada. Net interchange between U.S. and Canadian balancing authorities represents about one percent of total NWPP load.

The IntercontinentalExchange (ICE) has four trading points in the Northwest: Mid-Columbia (Mid-C), California-Oregon Border (COB), Nevada-Oregon Border (NOB) and Mona (Utah). Mid-C has the most traded volume by far, averaging more than 6,700 MW of daily on-peak physical trades in 2009. COB had almost 600 MW, NOB had 100 MW and Mona had 32 MW. Mid-C also has a fairly active physical forward market.

The Southwest electric market encompasses the Arizona, New Mexico, southern Nevada (AZ/NM/SNV) and the Rocky Mountain Power Area (RMPA) subregions of the Western Electric Coordinating Council (WECC). Peak demand is approximately 41 GW in summer and 29 GW in winter. There is 52 GW of generation capacity, composed mostly of gas and coal units.

The Southwest relies on nuclear and coal generators for baseload electricity, with gas units used as peaking resources. The coal generators are generally located in close proximity to coal mines, resulting in low delivered fuel costs. Some generation is jointly owned among multiple nearby utilities, including the Palo Verde nuclear plant, a 4,000-MW unit, which has owners in California and the Southwest.

The AZ/NM/SNV region is summer-peaking and experiences high loads due to air conditioning demand. The daily high temperatures average above 100 degrees in June through August in Phoenix. However, power prices tend to be the highest when there is also hot weather in Southern California, creating competition for the generation resources.

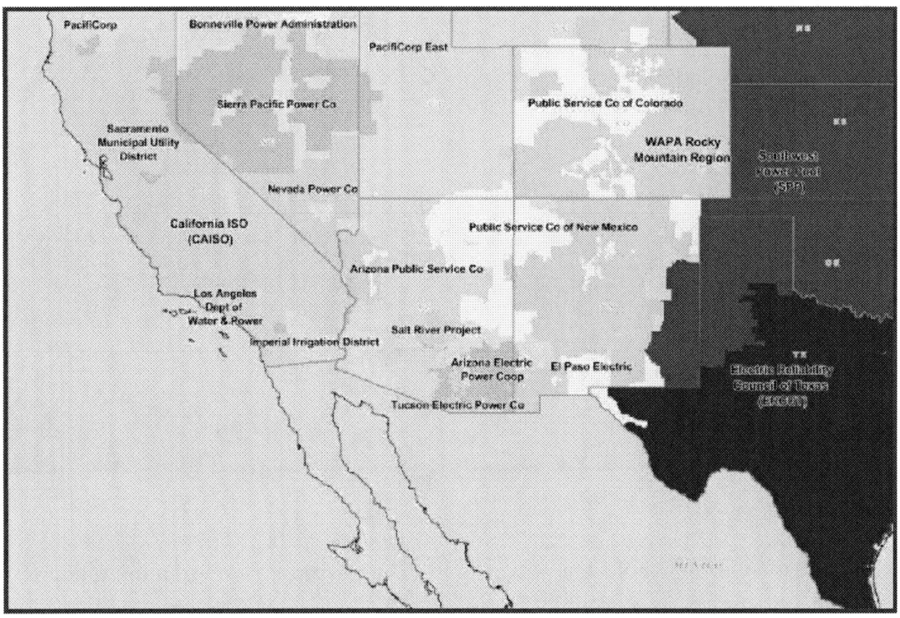

Southwest Electric Regions.

CAISO - California Independent System Operator

Market Profile

Geographic Scope

CAISO is a California nonprofit public benefit corporation started in 1998 when the state restructured its electric industry. The CAISO manages wholesale electricity markets, centrally dispatching electricity generation and facilities. In managing the grid, CAISO provides open access to the transmission system and performs long-term transmission planning. It manages energy and ancillary markets in day-ahead and real-time markets and is responsible for regional reliability.

Peak Demand

CAISO's all-time peak load was 50 GW in summer 2006.

Import and Exports

About 25 percent of CAISO's energy is supplied by imports, principally from two primary sources: the Southwest (Arizona, Nevada and New Mexico) and the Pacific Northwest (Oregon, Washington and British Columbia). Imports from the Pacific Northwest generally increase in the late spring when hydroelectric production peaks from increases in winter snowmelt and runoff.

Market Participants

CAISO's market participants include generators, retail marketers and utility customers, ranging from the three big investor-owned utilities (IOUs), which include Pacific Gas El Electric (PGElE), Southern California Edison (SCE) and San Diego Gas ElElectric (SDGElE), to municipalities and financial participants.

Membership and Governance

The CAISO has a board of governors that consists of five members appointed by the governor and confirmed by the California Senate. The board's role is to provide corporate direction, review and approve management's annual strategic plans and approve CAISO's operating and capital budgets. CAISO uses an informal stakeholder process to propose solutions to problems that may ultimately require a filing at FERC. Unlike other RTOs, which have a formal committee structure, CAISO's stakeholder process generally consists of rounds of dialogue with stakeholders on major policy issues.

Transmission

Owners

The Participating Transmission Owners (PTOs) in the CAISO control area include:

- Pacific Gas and Electric Co.,
- Southern California Edison,
- San Diego Gas and Electric, and
- Municipalities such as Vernon, Anaheim and Riverside.

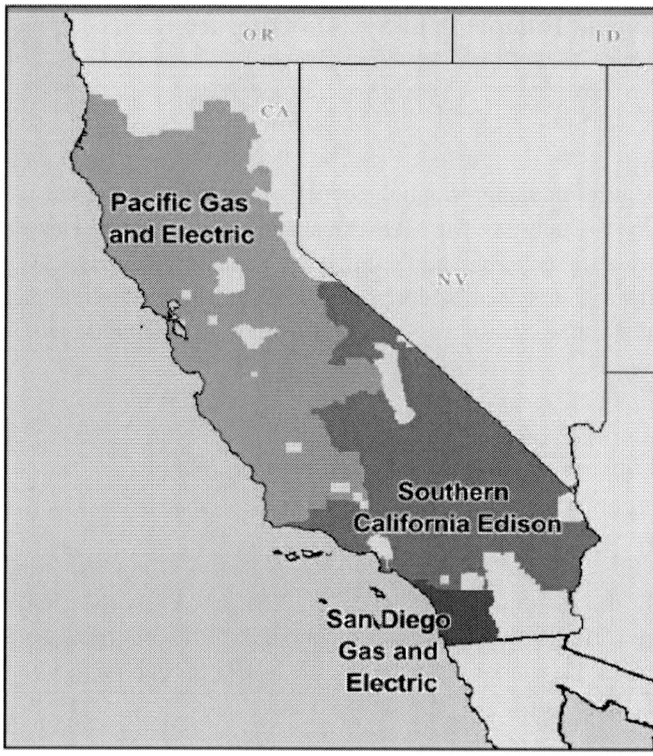

California Independent System Operator (CAISO).

Chronic Constraints

Areas of the system that are chronically constrained include the Humboldt region in the northwest corner of the state, import lines from the Southwest and Southern California (including San Diego).

Transmission Planning

CAISO conducts an annual transmission planning process with stakeholders that includes both short-term and longterm projects.

Supply Resource

Generating Mix

By plant capacity, the generating mix includes these sources:

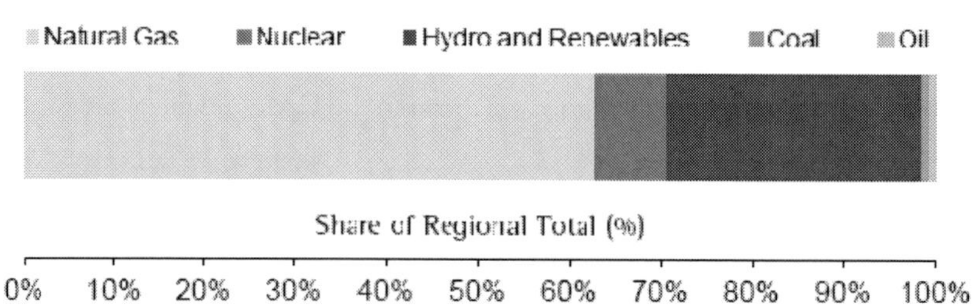

Demand Response

Demand-resource participation in the wholesale energy market is currently limited to a small amount of demand associated with water pumping loads. However, the market allows end-use loads that can be curtailed when directly dispatched in the real-time market to participate in the real-time energy and ancillary service (nonspinning reserve) markets. The California Public Utilities Commission is considering rules for allowing more retail demand-resource participation in the CAISO market.

Other demand response in California consists of programs for managing peak summer demands developed by the state's three major investor owned utilities. These demand-response programs are triggered based on criteria that are internal to the utility and not necessarily tied to market prices, although in early 2012 CAISO had a proposal pending at the Commission to integrate these reliability-based retail programs into its wholesale market.

Market Features and Functions

Energy Markets

Day-Ahead Market

The day-ahead market allows participants to secure prices for electric energy the day before the operating day and hedge against price fluctuations that can occur in real time. One day ahead of actual dispatch, participants submit supply offers and demand bids for energy. These bids are applied to each hour of the day and for each pricing location on the system.

From the offers and bids, CAISO constructs aggregate supply and demand curves for each location. The intersection of these curves identifies the market-clearing price at each location for every hour. Supply offers below and demand bids above the identified price are said to clear, meaning they are scheduled for dispatch. Offers and bids that clear are entered into a pricing software system along with binding transmission constraints to produce the locational marginal prices (LMP) for all locations.

Generator offers scheduled in the day-ahead settlement are paid the day-ahead LMP for the megawatts accepted. Scheduled suppliers must produce the committed quantity during real-time or buy power from the real-time market to replace what was not produced.

Likewise, wholesale buyers of electricity whose bids clear in the day-ahead market settlement pay for and lock in their right to consume the cleared quantity at the day-ahead LMP. Electricity use in real time that exceeds the day-ahead purchase is paid for at the real-time LMP.

Hour-Ahead Market

CAISO has an hour-ahead market for buying and selling imports and exports that it calls the hour-ahead scheduling process (HASP). HASP is designed to allow the ISO to re-optimize the market, given changes to internal supply and demand after the close of the day-ahead market.

In HASP, imports and exports between the ISO and neighboring regions are pre-dispatched 45 minutes before the start of each operating hour. These imports and exports are scheduled at a fixed level for the entire hour. However, resources within the CAISO and dynamic resources in neighboring regions (or balancing authority areas) can be dispatched every five minutes within each operating hour to meet real-time loads. A dynamic resource is a resource that is physically located in a neighboring region or balancing authority areas, yet is controllable by the CAISO market.

Real-Time Market

CAISO must coordinate the dispatch of generation and demand resources to meet the instantaneous demand for electricity. While the day-ahead energy market produces the schedule and financial terms of energy production and use for the operating day, a number of factors can change that schedule. Thus, to meet energy needs within each hour of the current day the CAISO operates a spot market for energy called the real-time market.

The real-time market uses final day-ahead schedules for resources within the ISO and final hour-ahead schedules for imports and exports as a starting point. It then re-dispatches resources every five minutes to balance generation and loads.

Prices resulting from the real-time market are only applicable to incremental adjustments to each resource's day-ahead schedule. Real-time bids can be submitted up to 75 minutes before the start of the operating hour.

Ancillary and Other Services

Ancillary services are those functions performed by electric generating, transmission, and system-control equipment to support the reliability of the transmission system. RTOs procure or direct the supply of ancillary services.

CAISO procures four ancillary services in the day-ahead and real-time markets:

- Regulation up: Units providing regulation up must be able to move quickly above their scheduled operating point in response to automated signals from the ISO to maintain the frequency on the system by balancing generation and demand.
- Regulation down: Units providing regulation down must be able to move quickly below their scheduled operating point in response to automated signals from the ISO.
- Spinning reserve: Resources providing spinning reserves must be synchronized with the grid (online, or spinning) and be able to respond within 10 minutes. This is more reliable than nonspinning reserves because it is already online and sychronized.
- Nonspinning reserve: Resources providing nonspinning reserves must be able to synchronize with the grid and respond within 10 minutes.

Regulation up and regulation down are used continually to maintain system frequency by balancing generation and demand. Spinning and nonspinning resources are used to maintain system frequency and stability during emergency operating conditions (such as unplanned outage of generation or transmission facilities) and major unexpected variations in load. Spinning and nonspinning resources are often referred to collectively as operating reserves.

Capacity Markets

Capacity markets provide a means for load-serving entities (LSEs) to procure capacity needed to meet forecast load, or resource adequacy (RA) requirements, and to allow generators to recover a portion of their fixed costs. They also provide economic incentives to attract investment in new and existing supply-side and demand-side capacity resources needed to maintain bulk power system reliability requirements.

The CAISO does not operate a formal capacity market, but it does have a mandatory RA requirement. The program requires that LSEs procure 115 percent of their aggregate system load on a monthly basis, unless a different reserve margin is mandated by the LSE's local regulatory authority. The program provides deliverability criteria each LSE must meet, as well as system and local capacity requirements.

Resources counted for RA purposes must make themselves available to the CAISO day-ahead and real-time markets for the capacity for which they were counted.

Market Power Mitigation

In electric power markets, mainly because of the largely non-storable nature of electricity and the existence of transmission constraints that can limit the availability of multiple suppliers to discipline market prices, some sellers from time to time have the ability to raise market prices. Market power mitigation is a market design mechanism to ensure competitive offers even when competitive conditions are not present.

Market power may need to be mitigated systemwide or locally when the exercise of market power may be particularly a concern for a local area. For example, when a transmission constraint creates the potential for local market power, the RTO may apply a set of behavioral and market outcome tests to determine if the local market is competitive and if generator offers should be adjusted to approximate price levels that would be seen in a competitive market – close to short-run marginal costs.

Reliability Must-Run

A reliability must-run (RMR) contract acts as an insurance policy, assuring that the CAISO has dispatch rights in order to reliably serve load in local import constrained areas. RMR contracts also help to mitigate any local market power that one or more units may have. The amount of generation capacity under RMR contracts dropped when local RA requirements were introduced. With more local resources being procured through RA contracts, the CAISO was able to significantly decrease its RMR designations in much of the system. Remaining generators with RMR contracts are located primarily in San Francisco and San Diego.

Financial Transmission Rights

As mentioned above, financial transmission rights (FTRs) give market participants an offset or hedge against transmission congestion costs in the day-ahead market. An FTR is a financial contract protecting the holder from costs arising from transmission congestion over a path or a sourceand-sink pair of locations on the grid. An FTR provides the holder with revenue, or charges, equal to the difference in congestion prices in the day-ahead market across the specific FTR transmission path. FTRs were originally formulated to protect LSEs from price uncertainty while redistributing excess congestion charges due to constrained conditions.

A related product is an auction revenue right (ARR). ARRs provide the holders with an upfront portion of the money raised in the FTR auctions. In general, they are allocated based on historical load served and, in some RTOs, can be converted to FTRs. As with FTRs, ARRs give transmission owners and eligible transmission service customers an offset or hedge against transmission congestion costs in the day-ahead market.

FTRs in California are referred to as congestion revenue rights (CRRs). Other than the name, the products are identical. CRRs are monthly or quarterly products. CRRs can be bought at auction or allocated by CAISO. Allocated CRRs receive the congestion value for a specific path, similar to a converted FTR. CAISO also allocates open market CRR auction revenue to LSEs based on their physical participation in the market, similar to an ARR in other markets. Given that both allocated CRRs and allocated auction revenues are based on physical market presences, LSEs often receive both. Finally, CRR revenue insufficiency is not possible as LSEs will be charged uplift if any shortfall is present.

Virtual Transactions

CAISO implemented convergence bidding, or virtual bidding, on Feb. 1, 2011. With the virtual bidding market feature, market participants can take financial positions in the day-ahead market that are liquidated in the real-time market. Virtual bidding is unique in CAISO because virtual positions taken on the export and import interties settle against the hour-ahead price, while internal virtual positions settle against the real-time dispatch price. Market participants can engage in either virtual demand or virtual supply transactions. A virtual demand transaction is a bid to buy at the day-ahead price and offer to sell at the real-time price. A virtual supply transaction is a bid to sell at the day-ahead price and buy at the real-time price.

Virtual supply and virtual demand may be submitted at any eligible pricing node in the CAISO system and there is no requirement for physical generation or load. The financial outcome for a particular participant is determined by the difference between the hourly day-ahead and real-time LMPs at the location at which the offer or bid clears. Thus, through this financial arbitrage opportunity, virtual transactions in theory help to narrow the difference between the day-ahead and real-time prices.

Credit Requirements

Credit requirements are important in organized electricity markets in which RTOs must balance the need for market liquidity against corresponding risk of default. Defaults within these markets are particularly troubling because losses due to default are spread among all market participants. Thus, each RTO's tariff specifies credit rules needed to participate in the markets. These requirements provide for credit evaluations, credit limits, allowed forms of collateral and the consequences of violations or defaults.

Settlements

RTOs must invoice market participants for their involvement in their markets. Settlements is the process by which the RTO determines the amounts to be paid associated with buying and selling energy, capacity and ancillary services, and paying administrative charges.

The CAISO calculates, accounts for and settles all charges and payments based on received settlement quality meter data. The CAISO settles the following charges: grid management charge, bid cost recovery, energy and ancillary services, CRR charges and payments, among others. The CAISO settles for three periods: the day-ahead market, the HASP and the real-time markets.

ISO-NE - New England Independent System Operator

Market Profile

Geographic Scope

As the RTO for New England, ISO-NE is responsible for operating wholesale power markets that trade electricity, capacity, transmission congestion contracts and related products, in addition to administering auctions for the sale of capacity. ISO-NE operates New England's high-voltage transmission network and performs long-term planning for the New England system. ISO-NE serves six New England states: Connecticut, Maine, Massachusetts, New Hampshire, Rhode Island and Vermont.

Peak Demand

New England's all-time peak load was 28 GW in 2006.

Import and Exports

ISO-NE is interconnected with the New York Independent System Operator (NYISO), TransEnergie (Québec) and the New Brunswick System Operator.

ISO-NE imports 12 percent of its annual energy needs from Québec. ISO-NE imports energy from and exports energy to NYISO.

New England receives imports from Québec and New Brunswick in most hours. Between New England and New York, power flows in alternate directions depending on market conditions.

Market Participants

The New England Power Pool (NEPOOL) consists of six sectors: (1) end-user sector; (2) publicly owned entities; (3) supplier sector; (4) transmission sector; (5) generation sector; and (6) alternative resources.

Membership and Governance

ISO-NE is a not-for-profit entity governed by a 10-member, independent, nonstakeholder board of directors. The sitting members of the board elect people to fill board vacancies.

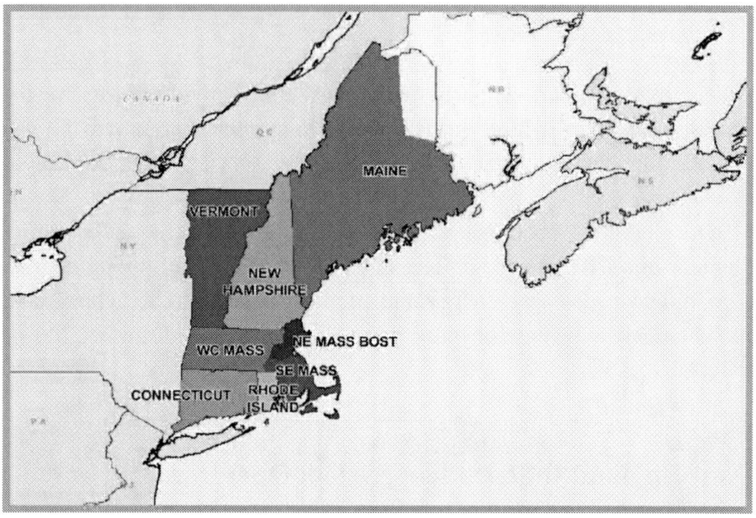

Independent System Operator of New England (ISO-NE).

Transmission

Owners

ISO-NE's transmission owners include:

- Bangor Hydro-Electric Co.,
- Central Maine Power Co.,
- New England Power Co.,
- Northeast Utilities System Cos.,
- NSTAR Electric Co.,
- Transmission Sector Provisional Group Member,
- The United Illuminating Co., and
- Vermont Electric Power Co. Inc.

Chronic Constraints

In 2009, New England completed a series of major transmission projects to improve reliability, including projects serving Boston, southwestern Connecticut and southeastern Massachusetts.

Transmission Planning

Each year, ISO-NE prepares a comprehensive 10-year regional system plan (RSP) that reports on the results of ISO system planning processes. Each plan includes forecasts of future loads (i.e., the demand for electricity measured in megawatts) and addresses how this demand may be satisfyed by adding supply resources; demand resources, including demand response and energy effi ciency; and new or upgraded transmission facilities. Each year's plan summarizes New England needs, as well as the needs in specific areas, and includes solutions and processes required to ensure the reliable and economic performance of the New England power system.

Supply Resources

Generating Mix

By plant capacity, the generating mix includes these sources:

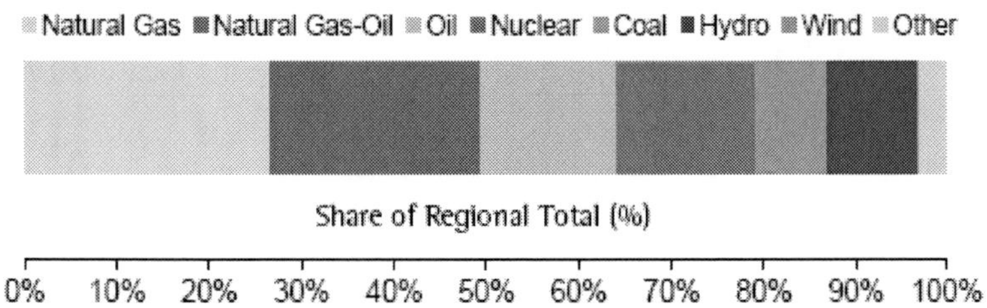

Demand Response

Currently, ISO-NE administers fi ve load-response programs for the New England wholesale electricity market. These include:

- **Real-Time 30-Minute Demand-Response Program:** These resources are required to respond within 30 minutes of the ISO's instructions.
- **Real-Time 2-Hour Demand Response Program:** This program requires demand resources to respond within two hours of the ISO's instructions.
- **Real-Time Profiled-Response Program:** These resources may be interrupted for anticipated capacity defi ciencies within a specifi ed time period and receive payment for a minimum of two hours.
- **Real-Time Price-Response Program:** These resources may interrupt (but are not required to do so) when they receive notice on the previous day. If they interrupt, they receive payment for the eligibility period.
- **Day-Ahead Load-Response Program**: An optional program that allows a participant in any of the realtime programs to offer interruptions concurrent with the day-ahead energy market. The participant is paid the day-ahead LMP for the cleared interruptions, and real-time deviations are charged or credited at the realtime LMP.

Market Features and Functions

Energy Markets

Day-Ahead

The day-ahead energy market allows market participants to secure prices for electric energy the day before the operating day and hedge against price fluctuations that can occur in real time. One day ahead of actual dispatch, participants submit supply offers and demand bids for energy. These bids are applied to each hour of the day and for each pricing location on the system.

In the day-ahead energy market, incremental offers and decremental bids (virtual supply offers and demand bids) can also be submitted, which indicate prices at which supply or demand are willing to increase or decrease their injection or withdrawal on the system. These INCs and DECs are tools market participants can use to hedge their positions in the day-ahead energy market.

From the offers and bids, the RTO constructs aggregate supply and demand curves for each location. The intersection of these curves identifies the market-clearing price at each location for every hour. Supply offers below and demand bids above the identified price are cleared and are scheduled. Offers and bids that clear are entered into a pricing software system along with binding transmission constraints to produce the locational marginal prices (LMPs) for all locations.

Hour-Ahead

None.

Real-Time

ISO-NE must coordinate the dispatch of generation and demand resources to meet the instantaneous demand for electricity. Supply or demand for the operating day can change for a variety of reasons, including unforeseen generator or transmission outages, transmission constraints or changes from the expected demand. While the day-ahead energy market produces the schedule and financial terms of energy production and use for the operating day, a number of factors can change that schedule. Thus, ISO-NE operates a spot market for energy, the real-time energy market, to meet energy needs within each hour of the current day.

ISO-NE clears the real-time energy market using supply offers, real-time load and offers and bids to sell or buy energy over the external interfaces. For generators, the market provides additional opportunities to offer supply to help meet incremental supply needs. Load-serving entities (LSEs) whose actual demand comes in higher than that scheduled in the day-ahead energy market may secure additional energy from the real-time energy market.

The real-time energy market financially settles the differences between the day-ahead scheduled amounts of load and generation and the actual real-time load and generation. Differences from the day-ahead quantities cleared are settled at the real-time LMP.

In real time, ISO-NE will issue dispatch rates and dispatch targets. These are five-minute price and megawatt signals based on the aggregate offers of generators, which will produce the required energy production. Market participants are, throughout the day, allowed to offer imports or request exports of electricity from neighboring control areas with at least one hour's notice.

Must-Offer Requirements

Market rules in RTOs include must-offer requirements for certain categories of resources for which withholding, a form of the exercise of market power, may be a concern. Where such rules apply, sellers must commit, or offer, the generators, and schedule and operate the facilities, into the applicable market.

Ancillary and Other Services

Ancillary services are those functions performed by electric generating, transmission and system-control equipment to support the transmission of electric power from generating resources to load while maintaining the reliability of the transmission system. RTOs procure or direct the supply of ancillary services.

ISO-NE procures ancillary services via the forward reserve market and its regulation market. The forward reserve market compensates generators for making available their unloaded operating capacity that can be converted into electric energy within 10 or 30 minutes when needed to meet system contingencies, such as unexpected outages. The Regulation Market compensates resources that ISO-NE instructs to increase or decrease output moment by moment to balance the variations in demand and system frequency to meet industry standards. The specific ancillary services ISO-NE procures in its markets include the following:

- Ten-Minute Spinning Reserves: provided by resources already synchronized to the grid and able to generate electricity within 10 minutes.
- Ten-Minute Nonspinning Reserves: provided by resources not currently synchronized to the grid but capable of starting and providing output within 10 minutes.
- Thirty-Minute Nonspinning Reserves: provided by resources not currently synchronized to the grid but capable of starting and providing output within 30 minutes.
- Regulation: provided by specially equipped resources with the capability to increase or decrease their generation output every four seconds in response to signals they receive from ISO-NE to control slight changes on the system.

Specialized ancillary services that are not bought and sold in these ancillary service markets include voltage support and black-start capability. Voltage support allows the New England control area to maintain transmission voltages. Black-start capability is the ability of a generating unit to go from a shutdown condition to an operating condition and start delivering power without assistance from a power system. ISO-NE procures these services via cost-based rates.

Capacity Markets

In ISO-NE's annual forward capacity auctions (FCA), both generator and demand resources offer capacity three years in advance of the period for which capacity will be supplied. The three-year lead time is intended to encourage participation by new resources and allow the market to adapt to resources leaving the market. Resources whose capacity clears the FCA acquire capacity supply obligations (CSOs). ISO-NE held its first two FCAs in 2008 for the 2010-11 and 2011-12 delivery years. The first full year of capacity market commitments began on June 1, 2010.

A third auction was held in December 2009 for the 2011-13 delivery years. The FCA process includes the modeling of transmission constraints to determine if load zones will be import- or export-constrained.

Market Power Mitigation

In ISO-NE, mitigation may be applied for physical withholding, economic withholding, uneconomic production, virtual transactions or other conduct if the conduct has a material effect on prices or uplift payments. The market monitor uses defined thresholds to identify physical and economic withholding and uneconomic generation, as well as defined thresholds to determine whether bids and offers would, if not mitigated, cause a material effect on LMPs or uplift charges.

Reliability Must-Run

None.

Financial Transmission Rights

New England FTRs are monthly and annual products. ISO-NE holds FTR auctions and then allocates the auction revenue to LSEs based on historical load. ISO-NE is the only RTO to settle accounts weekly but revenue insufficiency is possible. Month-to-month surplus is carried over and used to fund any deficiencies, and the true-up period for accounts occurs once a year.

Virtual Transactions

In ISO-NE, any market participant may submit INCs or DECs into the day-ahead market.

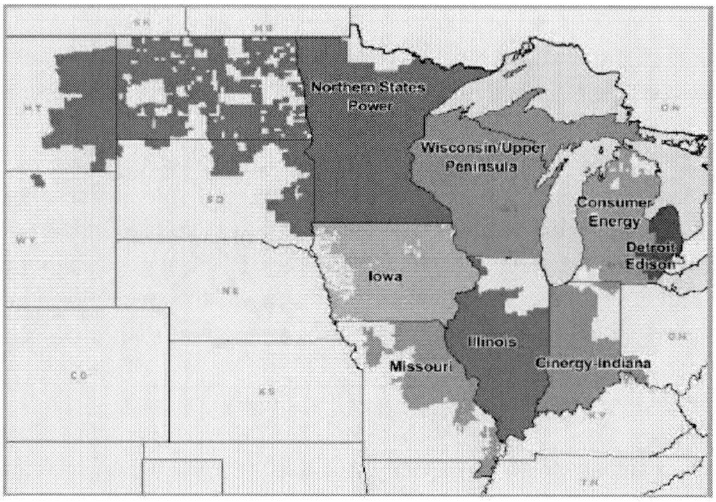

Midwest Independent System Operator (MISO).

MISO - Midwest Independent System Operator

Market Profile

Geographic Scope

MISO operates the transmission system and a centrally dispatched market in portions of 13 states in the Midwest, extending from western Pennsylvania to eastern Montana and from the Canadian border to the southern extremes of Illinois and Missouri. The system is operated from a primary control center in Carmel, Ind., and a second control center in St. Paul, Minn., for the western region.

MISO also serves as the reliability coordinator for additional systems outside of its market area, primarily to the north and northwest of the market footprint. MISO was not a power pool before organizing as an ISO in December 2001.

It began market operations in April 2005. In January 2009, MISO started operating an ancillary services market and combined its 24 separate balancing areas into a single balancing area.

Demand

MISO's peak demand was 116 GW in 2006.

Import and Exports

MISO has interconnections with the PJM and Southwest Power Pool (SPP) RTOs. It is also directly connected to TVA, the Western Area Power Administration and the electric systems of Manitoba and Ontario, plus several smaller systems.

MISO is a net importer of power overall, but the interchange with some areas can flow in either direction, depending on the relative loads and prices in the adjoining regions. Manitoba Hydro supplies a large part of MISO's load with its excess capacity, particularly in the summer.

Market Participants

MISO includes 34 transmission owners, whose assets define the MISO market area. MISO's market participants include generators, power marketers, transmission-dependent utilities and load-serving entities.

Membership and Governance

An independent board of directors of eight members, including the president, governs MISO. Directors are elected by the MISO membership from candidates provided by the board. An advisory committee of the membership provides advice to the board and information to the MISO stakeholders.

Membership includes entities with an interest in MISO's operation, such as state regulators and consumer advocates, as well as transmission owners, independent power producers, power marketers and brokers, municipal and cooperative utilities and large-volume customers.

Transmission

Owners

The transmission owners in MISO include:

- Alliant Energy
- American Transmission Co.
- Ameren (Missouri and Illinois)
- American Transmission Systems

- Cinergy Services (Duke)
- Indianapolis Power and Light
- ITC
- Michigan Public Power Agency
- NSP Companies (Xcel)
- Northern Indiana Public Service Co.
- Otter Tail Power
- MidAmerican Energy

Chronic Constraints

MISO has certain pathways that are more likely to become congested, but the likelihood and pattern of congestion in any area is subject to weather patterns, wind production and interchange with external regions. When load is high in the eastern part of MISO and to the east in PJM, constraints occur on pathways from the Minnesota and Wisconsin areas through Chicago and across Indiana and Ohio. A particular congestion point with this pattern is northern Indiana. When colder weather hits Minnesota and the Dakotas, there is often congestion in the northern direction, particularly in Iowa. Higher wind production can cause localized constraints in some areas and can cause congestion in pathways from southern Minnesota and western Iowa moving eastward.

Transmission Planning

The main vehicle MISO uses for transmission planning is the Midwest ISO Transmission Expansion Plan developed by the MISO planning department in collaboration with transmission owners and other stakeholders who form the planning advisory committee. The plan is for two years. Once approved by the board, the plan becomes the responsibility of the transmission owners.

Supply Resources

Generating Mix

By plant capacity, the generating mix includes these sources:

Demand Response

Demand-side resources are able to participate in MISO's markets in providing capacity, energy in both the day-ahead and real-time markets and ancillary services.

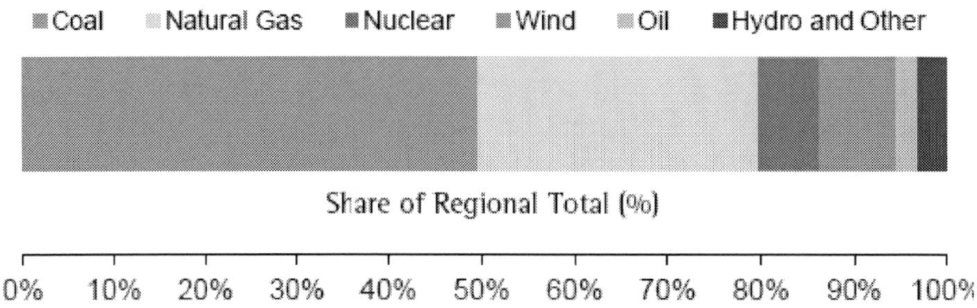

Market Features and Functions

Energy Markets

Day-Ahead Market

The day-ahead market allows market participants to secure prices for electric energy the day before the operating day and hedge against price fluctuations that can occur in real time. One day ahead of actual dispatch, participants submit supply offers and demand bids for energy. These bids are applied to each hour of the day and for each pricing location on the system.

In the day-ahead market, incremental offers and decremental bids (virtual supply offers and demand bids) can also be submitted, although they are not associated with physical resources or actual load. These INCs and DECs are tools market participants can use to hedge their real time commitments or to arbitrage the day-ahead to real-time price spread.

From the offers and bids, the RTO constructs aggregate supply and demand curves for each location. The intersection of these curves identifies the market-clearing price at each location for every hour. Supply offers below and demand bids above the identified price are scheduled. Offers and bids that clear are entered into a pricing software system along with binding transmission constraints to produce the locational marginal prices (LMPs) for all locations.

Generators and offers scheduled in the day-ahead settlement are paid the day-ahead LMP for the megawatts accepted. Scheduled suppliers must produce the committed quantity during real-time or buy power from the real-time marketplace to replace what was not produced.

Likewise, wholesale buyers of electricity and virtual demand whose bids to buy clear in the day-ahead market settlement pay for and lock in their right to consume the cleared quantity at the day-ahead LMP. Electricity use in real-time that exceeds the day-ahead purchase is paid for at the real-time LMP.

Hour-Ahead Market

Not applicable for MISO.

Real-Time Market

MISO must coordinate the dispatch of generation and demand resources to meet the instantaneous demand for electricity. Supply or demand for the operating day can change for a variety of reasons, including unforeseen generator or transmission outages, transmission constraints or changes from the expected demand. While the day-ahead energy market produces the schedule and financial terms for the bulk of the physical transactions, a number of factors usually change the day-ahead result. Thus, MISO operates a spot market for energy, the real-time energy market, to meet actual energy needs within each hour of the operating day.

The real-time market is prepared for at the conclusion of the day-ahead market on the day before the operating day. MISO clears the real-time energy market using supply offers, real-time load and external offers. For generators, the market provides additional opportunities to offer supply to help meet incremental needs. LSEs whose actual demand comes in higher than what was scheduled in the day-ahead market may secure additional energy from the real-time market.

The real-time energy market financially settles the differences between the day-ahead scheduled amounts of load and generation and the actual real-time load and generation. Participants either pay or are paid the real-time LMP for the amount of load or generation in megawatt-hours that deviates from their day-ahead schedule.

In real-time, MISO issues dispatch rates and dispatch targets. These are five-minute price and megawatt signals based on the aggregate offers of generators, which will produce the required energy production. Market participants are, throughout the day, allowed to offer imports or request exports of electricity from neighboring control areas by submitting transmission schedules into or out of MISO.

In real-time, generators can also deviate from the day-ahead clearing schedule by self-scheduling, which means that MISO will run a given unit without regard to the unit's economics unless running the unit presents a reliability concern.

During the operating day, the real-time market acts as a balancing market for load with physical resources used to meet that load. A market price for energy and for each of the ancillary services is calculated for each five-minute dispatch interval and the resulting five-minute prices are rolled into hourly prices for billing and payment. Differences in the real-time operation from the day-ahead clearing, including all virtual transactions, are settled at the real-time price.

Must-Offer Requirements

Market rules in RTOs include must-offer requirements for certain categories of resources for which withholding, which could be an exercise of market power, may be a concern. Where such rules apply, sellers must commit, or offer, the generators, and schedule and operate the facilities, in the applicable market.

In MISO, generators who supply capacity to meet the RTO resource adequacy requirement for load are required to offer into the day-ahead and real-time markets for energy and the ancillary services for which they are qualified.

Ancillary and Other Services

Ancillary services are those functions performed by electric generating, transmission and system-control equipment to support the transmission of electric power from generating resources to load while maintaining the reliability of the transmission system. RTOs procure or direct the supply of ancillary services.

MISO procures ancillary services via the co-optimized energy and ancillary services market and includes the following services:

- Spinning Reserves: provided by resources already synchronized to the grid and able to provide output within 10 minutes.
- Supplemental (nonspinning) Reserves: provided by resources not currently synchronized to the grid but capable of starting and providing output within 10 minutes.
- Regulation: provided by specially equipped resources with the capability to increase or decrease their generation output every four seconds in response to signals they receive to control slight changes on the system.

Capacity Markets

Capacity markets are a construct to provide assurance to government entities and to NERC a means for load-serving entities (LSEs) to prove they have procured capacity needed to meet forecast load and to allow generators to recover a portion of their fixed costs. They also provide economic incentives to attract investment in new and existing supply-side and demand-side capacity resources.

MISO maintains a monthly capacity requirement on all LSEs based on the load forecast plus reserves. LSEs are required to specify to MISO what physical capacity, including demand resources, they have designated to meet their load forecast. This capacity can be acquired either through bilateral purchase or self-supply. Additionally, MISO conducts a monthly auction to provide an opportunity for load that has not arranged all of its capacity to procure its needs from uncommitted resources.

Market Power Mitigation

In electric power markets, mainly because of the largely nonstorable nature of electricity and the existence of transmission constraints that can limit the availability of multiple suppliers to discipline market prices, some sellers from time to time have the ability to raise market prices. Market power mitigation is a market design mechanism to ensure competitive offers even when competitive conditions are not present.

Reliability Must-Run

None.

Financial Transmission Rights

Financial transmission rights (FTRs) give market participants an offset or hedge against transmission congestion costs in the day-ahead market. An FTR is a financial contract protecting the holder from costs arising from transmission congestion over a path or a source-and-sink pair of locations on the grid. An FTR provides the holder with revenue, or charges, equal to the difference in congestion prices in the day-ahead market across the specific FTR transmission path. FTRs were originally formulated to protect LSEs from price uncertainty while redistributing excess congestion charges due to constrained conditions. Other market participants such as financial-only participants may purchase FTRs through the RTO's auctions or through secondary market purchases.

MISO FTRs are monthly and annual products.

Virtual Transactions

A virtual transaction allows a participant to buy or sell power in the day-ahead market without requiring physical generation or load. Cleared virtual supply (increment or virtual offers, or INCs) in the day-ahead energy market at a particular location in a certain hour creates a financial obligation for the participant to buy back the bid quantity in the real-time market at that location in that hour. Cleared virtual demand (decrement or virtual bids, or DECs) in the day-ahead market creates a financial obligation to sell the bid quantity in the real-time market. The financial outcome for a particular participant is determined by the difference between the hourly day-ahead and real-time LMPs at the location at which the offer or bid clears. Thus, through this financial arbitrage opportunity, virtual transactions in theory help to narrow the difference between the day-ahead and real-time prices. MISO allows virtual bids and offers into its day-ahead energy market where the bids and offers are included in the determination of the LMP along with physical resource offers and actual load bids. Market participants whose virtual transactions clear in the day-ahead market, have their positions cleared in the real-time market at the real-time price. Virtual bids and offers are allowed in MISO at any pricing node or aggregate of pricing nodes.

Credit Requirements

Credit requirements are important in markets in which RTOs must balance the need for market liquidity against corresponding risk of default. Defaults within these markets are particularly troubling because losses due to default are borne among all market participants. Thus, each RTO's tariff specifies credit rules needed to participate in the markets. These requirements provide for credit evaluations, credit limits, allowed forms of collateral and the consequences of violations or defaults.

Settlements

RTOs must invoice market participants for their involvement in their markets. The RTO determines the amount owed associated with buying and selling energy, capacity and ancillary services and paying various administrative charges. Settlements for market activity in MISO are finalized seven days after the operating day and payable after 14 days.

NYISO - New York Independent System Operator

Market Profile

Geographic Scope

Prior to restructuring of the electric industry in the 1990s, New York's private utilities and public power authorities owned and operated New York's electric system. Operation of the electric grid was coordinated by a voluntary collaboration of the utilities and power authorities as the New York Power Pool (NYPP). The creation of the New York Independent

System Operator (NYISO) was authorized by FERC in 1998. The formal transfer of the NYPP's responsibilities to the NYISO took place on Dec. 1, 1999.

The NYISO footprint covers the entire state of New York. NYISO is responsible for operating wholesale power markets that trade electricity, capacity, transmission congestion contracts, and related products, in addition to administering auctions for the sale of capacity. NYISO operates New York's high-voltage transmission network and performs long-term planning.

Demand

NYISO's all-time peak load was 34 GW in 2006.

Imports and Exports

NYISO imports and exports energy through interconnections with ISO-NE, PJM, TransEnergie (Quebec) and Ontario.

Market Participants

NYISO's market participants include generators, transmission owners, financial institutions, traditional local utilities, electric co-ops and industrials.

Membership and Governance

NYISO is governed by an independent 10-member board of directors and management, business issues and operating committees. Each committee oversees its own set of working groups or subcommittees. These committees comprise transmission owners, generation owners and other suppliers, consumers, public power and environmental entities. Tariff revisions on market rules and operating procedures filed with the Commission are largely developed through consensus by these committees. The members of the board, as well as all employees, must not be directly associated with any market participant or stakeholder.

Transmission

Transmission Owners

NYISO's transmission owners include:

- Central Hudson Gas El Electric Corp.
- Consolidated Edison Co. of New York (ConEd)
- Long Island Power Authority (LIPA)
- New York Power Authority (NYPA)
- New York State Electric and Gas Corp. (NYSEG)
- National Grid
- Orange El Rockland Utilities
- Rochester Gas and Electric Corp.

Chronic Constraints

The chronic transmission constraints in NYISO are in the southeastern portion of the state, leading into New York City and Long Island. As a result of their dense populations,

New York City and Long Island are the largest consumers of electricity. Consequently, energy flows from the west and the north toward these two large markets, pushing transmission facilities near their operational limits. This results in transmission constraints in several key areas, often resulting in higher prices in the New York City and Long Island markets.

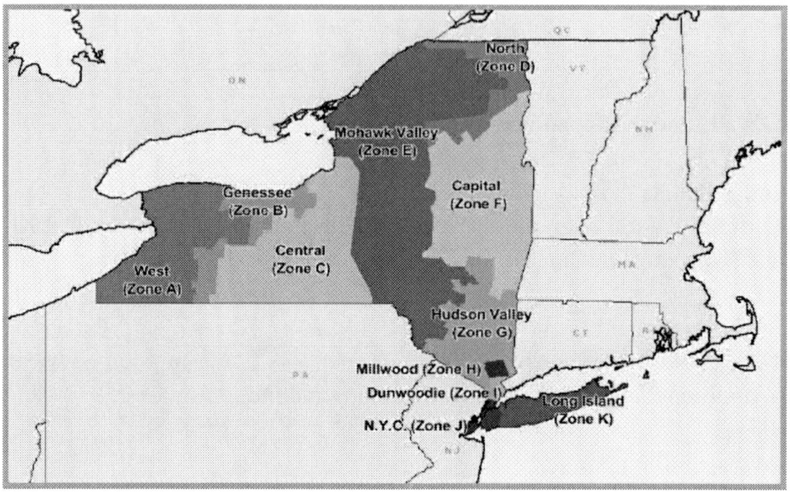

New York Independent System Operator (NYISO).

Supply Resources

Generating Mix

By plant capacity, the generating mix includes these sources:

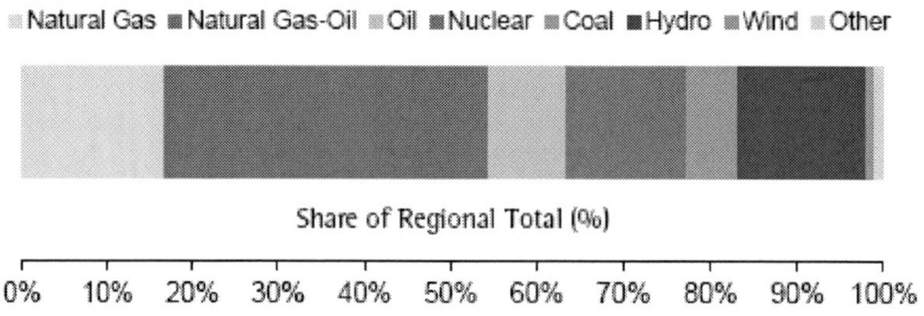

Demand Response

NYISO has four demand-response (DR) programs: the emergency demand-response program (EDRP), the installed capacity (ICAP) special case resources program (SCR), the day-ahead demand-response program and the demand-side ancillary services program.

Both the emergency and special cases programs can be deployed in energy shortage situations to maintain the reliability of the bulk power grid. Both programs are designed to reduce power usage by shutting down businesses and large power users. Companies, mostly industrial and commercial, sign up to take part in the programs. The companies are paid by NYISO for reducing energy consumption when asked to do so. Reductions are voluntary for

EDRP participants. SCR participants are required to reduce power usage and as part of their agreement are paid in advance for agreeing to cut power usage on request.

NYISO's day-ahead DR program allows energy users to bid their load reductions into the day-ahead market. Offers determined to be economic are paid at the market clearing price. Under day-ahead DR, flexible loads may effectively increase the amount of supply in the market and moderate prices.

The ancillary services program provides retail customers that can meet telemetry and other qualifications the ability to bid their load curtailment capability into the day-ahead market or real-time market to provide reserves and regulation service. Scheduled offers are paid the marketing clearing price for reserves or regulation.

Market Features and Functions

Energy Markets

Day-Ahead Market

The day-ahead market allows market participants to secure prices for electric energy the day before the operating day and hedge against price fluctuations that can occur in real time. One day ahead of actual dispatch, participants submit supply offers and demand bids for energy. These bids are applied to each hour of the day and for each pricing location on the system.

In the day-ahead market, virtual supply offers and demand bids can also be submitted. These are tools market participants can use to hedge their positions in the day-ahead market.

From the offers and bids, the RTO constructs aggregate supply and demand curves for each location. The intersection of these curves identifies the market-clearing price at each location for every hour. Supply offers below, and demand bids above, the identified price are scheduled. Offers and bids that clear are then entered into a pricing software system along with binding transmission constraints to produce the locational marginal prices (LMPs) for all locations. The NYISO refers to these as locational based marginal prices, or LBMPs.

Generators and offers scheduled in the day-ahead settlement are paid the day-ahead LBMP for the megawatts accepted. Scheduled suppliers must produce the committed quantity during real-time or buy power from the real-time marketplace to replace what was not produced.

Likewise, wholesale buyers of electricity and virtual demand whose bids to buy are accepted in the day-ahead market pay for and lock in their right to consume the cleared quantity at the day-ahead LBMP. Electricity used in real-time that exceeds the day-ahead purchase is paid for at the real-time LBMP.

Hour-Ahead Market

The hour-ahead market allows buyers and sellers of electricity to balance unexpected increases or decreases of electricity use after the day-ahead market closes. Bids and offers are submitted an hour ahead of time. Prices are set based on those bids and offers, generally for use in matching generation and load requirements, but those prices are advisory only. Hour-ahead scheduling is completed at least 45 minutes prior to the beginning of the dispatch hour after NYISO reviews transmission outages, the load forecast, reserve requirements and hour-ahead generation and firm transaction bids, among other things.

Real-Time Market

NYISO must coordinate the dispatch of generation and demand resources to meet the instantaneous demand for electricity. Supply or demand for the operating day can change for a variety of reasons, including unforeseen generator or transmission outages, transmission constraints or changes from the expected demand. While the day-ahead market produces the schedule and financial terms of energy production and use for the operating day, a number of factors can change that schedule. Thus, NYISO operates a spot market for energy, the real-time energy market, to meet energy needs within each hour of the current day.

Real-time energy market outcomes are based on supply offers, real-time load and offers and bids to sell or buy energy. LSEs whose actual demand comes in higher than that scheduled in the day-ahead market may secure additional energy from the real-time market. For generators, the market provides additional opportunities to offer supply to help meet additional needs.

The real-time energy market financially settles the differences between the day-ahead scheduled amounts of load and generation and the actual real-time load and generation. Those who were committed to produce in the day-ahead are compensated at (or pay) the real-time LBMP for the megawatts under- or over-produced in relation to the cleared amount. Those who paid for day-ahead megawatts are paid (or pay) the real-time LBMP for megawatts under- or over-consumed in real-time.

Real-time dispatch of generators occurs every five minutes, as does the setting of the real-time prices used for settlement purposes. Market participants may participate in the day-ahead, hour-ahead, and the real-time market.

Must-Offer Requirements

Under the NYISO capacity auction rules, entities that offer capacity into an auction that is subsequently purchased by load are required to offer that amount of capacity into the day-ahead energy market. This rule ensures that capacity sold through the capacity auctions is actually delivered into the market.

Ancillary and Other Services

Ancillary services are those functions performed by electric generating, transmission and system-control equipment to support the transmission of electric power from generating resources to load while maintaining the reliability of the transmission system. RTOs procure or direct the supply of ancillary services.

NYISO administers competitive markets for ancillary services that are required to support the power system. The two most important types of ancillary services are operating reserves and regulation. Operating reserves and regulation are typically provided by generators, but NYISO allows demand-side providers to participate in these markets as well. Operating reserve resources can either be spinning (online with additional ramping ability) or nonspinning (off-line, but able to start and synchronize quickly). NYISO relies on regulating resources that can quickly adjust their output or consumption in response to constantly changing load conditions to maintain system balance.

The NYISO relies on the following types of ancillary services:

- Ten-Minute spinning reserves: provided by resources already synchronized to the grid and able to provide output within 10 minutes.
- Ten-Minute nonspinning reserves: provided by resources not currently synchronized to the grid but capable of starting and providing output within 10 minutes.
- Thirty-Minute nonspinning reserves: provided by resources not currently synchronized to the grid but capable of starting and providing output within 30 minutes.
- Regulation: provided by resources with the capability to increase or decrease their generation output within seconds in order to control changes on the system.

Capacity Markets

Capacity markets provide a means for load-serving entities (LSEs) to procure capacity needed to meet forecast load and to allow generators to recover a portion of their fixed costs. They also provide economic incentives to attract investment in new and existing supply-side and demand-side capacity resources in New York as needed to maintain bulk power system reliability requirements.

In NYISO's capacity market, LSEs procure capacity through installed-capacity (ICAP) auctions, self-supply and bilateral arrangements based on their forecasted peak load plus a margin. The NYISO conducts auctions for three different service durations: the capability period auction (covering six months), the monthly auction and the spot market auction.

New York has capacity requirements for three zones: New York City, Long Island and New York-Rest of State. The resource requirements do not change in the monthly auctions and ICAP spot market auctions relative to the capability period auction. The shorter monthly auctions are designed to account for incremental changes in LSE's load forecasts.

Market Power Mitigation

In electric power markets, mainly because of the largely nonstorable nature of electricity and the existence of transmission constraints that can limit the availability of multiple suppliers to discipline market prices, some sellers from time to time have the ability to raise market prices. Market power mitigation is a market design mechanism to ensure competitive offers even when competitive conditions are not present.

Market power may need to be mitigated on a systemwide basis or on a local basis. When a transmission constraint creates the potential for local market power, the RTO may apply a set of behavioral and market outcome tests to determine if the local market is competitive and if generator offers should be adjusted to approximate price levels that would be seen in a competitive market.

The categories of conduct that may warrant mitigation by NYISO include physical withholding, economic withholding and uneconomic production by a generator or transmission facility to obtain benefits from a transmission constraint. Physical withholding is not offering to sell or schedule energy provided by a generator or transmission facility capable of serving a NYISO market. Physical withholding may include falsely declaring an outage, refusing to offer or schedule a generator or transmission facility; making an unjustifiable change to operating parameters of a generator that reduces its availability; or operating a generator in real-time at a lower output level than the generator would have been expected to produce had the generator followed NYISO's dispatch instructions. Economic withholding is submitting bids for a generator that are unjustifiably high so that the generator is not dispatched.

NYISO will not impose mitigation unless the conduct causes or contributes to a material change in prices, or substantially increases guarantee payments to participants.

Virtual bidding is subject to mitigation under certain circumstances as well. NYISO may limit the hourly quantities of virtual bids for supply or load that may be offered in a zone by a market participant whose virtual bidding practices are determined to contribute to an unwarranted divergence of LBMPs (location-based marginal prices) between the day-ahead and real-time markets. If the NYISO determines that the relationship between zonal LBMPs in a zone in the day-ahead market and the real-time market is not what would be expected under conditions of workable competition, and that the virtual bidding practices of one or more market participants has contributed to this divergence, then a mitigation measure may be imposed.

Price Caps

NYISO does not have price caps. It employs a bid cap of $1,000/MWh for its day-ahead and real-time markets.

ICAP for New York City is subject to offer caps and floors. Offer caps in New York City are based on reference levels or avoided costs. Capacity from generators within New York City must be offered in each ICAP spot market auction, unless that capacity has been exported out of New York or sold to meet ICAP requirements outside New York City.

Local Market Power Mitigation

Generators in New York City are subject to automated market power mitigation procedures because New York City is geographically separated from other parts of New York; plus, generators in New York City have been deemed to have market power.

These automated procedures determine whether any day-ahead or real-time energy bids, including start-up costs bids and minimum generation bids, but excluding ancillary services bids, exceed the tariff's thresholds for economic withholding, and, if so, determine whether such bids would cause material price effects or changes in guarantee payments. If these two tests are met, mitigation is imposed automatically.

For example, the threshold for economic withholding regarding energy and minimum generation bids is a 300 percent increase or an increase of $100/MWh over the applicable reference level, whichever is lower. Energy or minimum generation bids below $25/MWh are not considered economic withholding. Regarding operating reserves and regulation bids, a 300 percent increase or an increase of $50/MW over the applicable reference level, whichever is lower, is the threshold for determining whether economic withholding has occurred. In this instance, bids below $5/ MW are not considered economic withholding. If an entity's bids meet these thresholds, the applicable reference level is substituted for the entity's actual bid to determine the clearing price.

Reliability Must-Run

NYISO has no reliability must-run provisions.

Financial Transmission Rights

Financial transmission rights (FTRs) give market participants an offset, or hedge, against transmission congestion costs in the day-ahead market. The NYISO refers to FTRs as transmission congestion contracts (TCCs). Other than the name, FTRs and TCCs are essentially the same. An FTR is a financial contract protecting the holder from costs arising from transmission congestion over a path or a sourceand-sink pair of locations (or nodes) on the grid. An FTR provides the holder with revenue, or charges, equal to the difference in congestion prices in the day-ahead market across the specific FTR transmission path. FTRs were originally formulated to protect LSEs from price uncertainty. Other market participants, including financial-only participants, may purchase FTRs through the RTO's auctions.

A related product is an auction revenue right (ARR), which provides the holders with an upfront portion of the money raised in the TCC auctions. In general, they are allocated based on historical load served. As with FTRs, ARRs give transmission owners and eligible transmission service customers an offset, or hedge, against transmission congestion costs in the day-ahead market.

Virtual Transactions

A virtual transaction is a specific kind of transaction that allows a participant to buy or sell power in the day-ahead market without requiring physical generation or load.

Cleared virtual supply (virtual offers) in the day-ahead energy market at a particular location in a certain hour creates a financial obligation for the participant to buy back the bid quantity in the real-time market at that location in that hour. Cleared virtual demand (virtual bids) in the day-ahead market creates a financial obligation to sell the bid quantity in the real-time market. The financial outcome is determined by the difference between the hourly day-ahead and real-time LBMPs at the location at which the offer or bid clears. Thus, through this financial arbitrage opportunity, virtual transactions in theory help to narrow the difference between the day-ahead and real-time prices.

Virtual bidding in NYISO takes place on a zonal level, not a nodal level.

Credit Requirements

Credit requirements are important in organized electricity markets in which RTOs must balance the need for market liquidity against corresponding risk of default. Losses due to default are borne among all market participants. Thus, each RTO's tariff specifies credit rules needed to participate in the markets. These requirements provide for credit evaluations, credit limits, allowed forms of collateral and the consequences of violations or defaults.

Settlements

RTOs must invoice market participants for their involvement in their markets. Settlements is the process by which the RTO determines the amounts owed and to be paid associated with buying and selling energy, capacity, ancillary services and paying various administrative charges.

NYISO uses a two-settlement process for its energy markets. The first settlement is based on day-ahead bids and offers, which clear the market and are scheduled. The second settlement is based on the real-time bids and the corresponding real-time dispatch.

PJM - The PJM Interconnection

Market Profile

Geographic Scope

The PJM Interconnection operates a competitive wholesale electricity market and manages the reliability of its transmission grid. PJM provides open access to the transmission and performs long-term planning.

In managing the grid, PJM centrally dispatches generation and coordinates the movement of wholesale electricity in all or part of 13 states (Delaware, Illinois, Indiana, Kentucky, Maryland, Michigan, New Jersey, North Carolina, Ohio, Pennsylvania, Tennessee, Virginia and West Virginia) and the District of Columbia. PJM's markets include energy (day-ahead and real-time), capacity and ancillary services.

PJM was founded in 1927 as a power pool of three utilities serving customers in Pennsylvania and New Jersey.

In 1956, with the addition of two Maryland utilities, it became the Pennsylvania-New Jersey-Maryland Interconnection, or PJM. PJM became a fully functioning ISO in 1996 and, in 1997, it introduced markets with bid-based pricing and locational market pricing (LMP). PJM was designated an RTO in 2001.

Demand

PJM's all-time peak load of 159 GW occurred in summer 2006.

Imports and Exports

PJM has interconnections with Midwest ISO and New York ISO. PJM also has direct interconnections with the Tennessee Valley Authority (TVA), Progress Energy Carolinas and the Virginia and Carolinas Area (VACAR), among other systems.

PJM market participants import energy from, and export energy to, external regions continuously. At times, PJM is a net importer of electricity and, at other times, PJM is a net exporter of electricity.

Market Participants

PJM's market participants include power generators, transmission owners, electric distributors, power marketers, electric distributors and large consumers.

Membership and Governance

PJM has a two-tiered governance model consisting of a board of managers and the members committee. PJM is governed by a 10-member board, nine of whom PJM members elect. The board appoints the tenth, the president and CEO, to supervise day-to-day operations. The board is generally responsible for oversight of system reliability, operating efficiency and short and long-term planning.

The board ensures that no member or group of members exerts undue influence. The members committee, which advises the board, is composed of five voting sectors representing power generators, transmission owners, electric distributors, power marketers and large consumers.

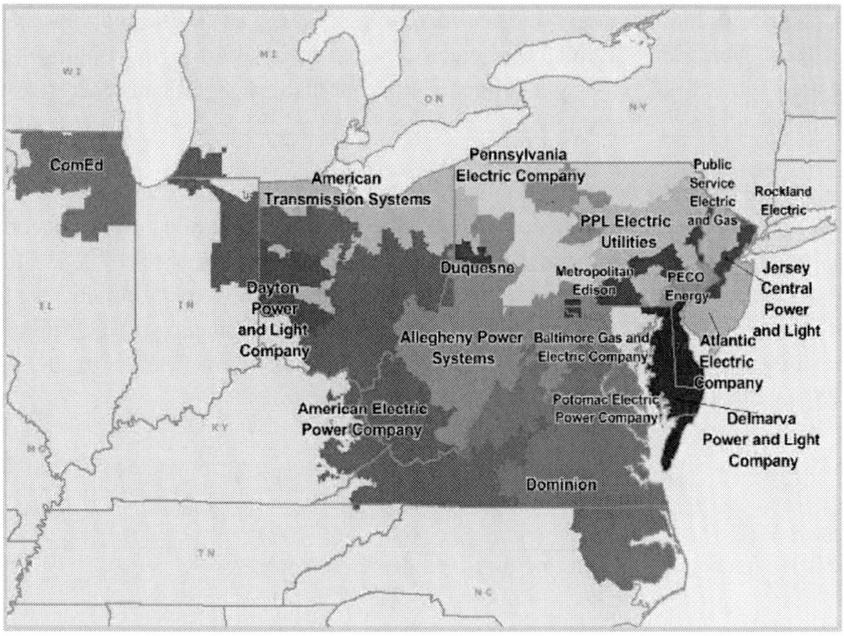

PJM Regional Transmission Organization.

Transmission

Transmission Owners

The largest transmission owners in PJM include:

- AEP,
- First Energy,
- PSE&G,
- Dominion,
- Philadelphia Electric, and
- Commonwealth Edison.

Chronic Constraints

In general, transmission paths extending from generation sources in western PJM to load centers in eastern PJM tend to become constrained, particularly during peak load conditions. PJM's Mid-Atlantic markets rely on generation in the western part of PJM due to the retirements of eastern units and the location of new generation capacity in western areas, such as western Pennsylvania, West Virginia and eastern Ohio. Eastern PJM relies on transmission across Pennsylvania and up from southwestern PJM to import power from sources west and southwest. Eastern PJM relies on transmission capability to replace retired generation and to meet demand growth.

Congestion on the eastern interface also constrains power flows from the District of Columbia, Baltimore and Northern Virginia to New Jersey, Delmarva Peninsula and Philadelphia load centers. The high-voltage, bulk power transmission pathway within portions of the states of Pennsylvania, West Virginia, Virginia and Maryland serve the densely populated load centers of the metropolitan areas of Baltimore, the District of Columbia and Northern Virginia. The electricity needs of Washington-BaltimoreNorthern Virginia are supplied not only by local generation but also by significant energy transfers to those areas.

Transmission Planning

PJM's regional transmission expansion plan (RTEP) identifies transmission system additions and improvements needed to keep electricity flowing within PJM. Studies are conducted to test the transmission system against national and regional reliability standards. These studies look forward to identify future transmission overloads, voltage limitations and other reliability standards violations. PJM then develops transmission plans to resolve violations that could otherwise lead to overloads and blackouts.

Supply Resources

Generation Mix

By plant capacity, the generating mix includes these sources:

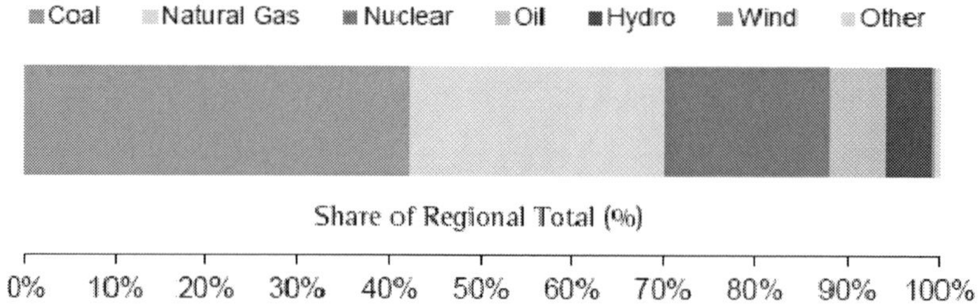

Demand Response

End-use customers providing demand response have the opportunity to participate in PJM's energy, capacity, synchronized reserve and regulation markets on an equal basis with generators. All demand-response programs can be grouped into emergency or economic programs. The emergency program compensates end-use customers who reduce their usage during emergency conditions on the PJM system. Participation in the emergency program may be voluntary or mandatory and payments may include energy payments, capacity payments or both. There are three options for emergency program registration and participation: energy only, capacity only and capacity-plus-energy.

The economic program allows end-use customers to reduce electricity consumption in the energy markets and receive a payment when LMPs are high. Under this program, all hours are eligible and all participation is voluntary. Participation in the program takes three forms: submitting a sell offer into the day-ahead market that clears; submitting a sell offer into the real-time market that is dispatched; and self-scheduling load reductions while providing notification to PJM. End-use customers participate in demand response in PJM through members called curtailment service providers, or CSPs, who act as agents for the customers. CSPs aggregate the demand of retail customers, register that demand with PJM, submit the verification of demand reductions for payment by PJM and receive the payment from PJM.

The payment is divided among the CSP and its retail customers based on private agreements between them.

Market Features and Functions

Energy Markets

Day-Ahead Market

The day-ahead market allows market participants to secure prices for electric energy the day before the operating day and hedge against price fluctuations that can occur in real-time. One day ahead of actual dispatch, participants submit supply offers and demand bids for energy. These bids are applied to each hour of the day and for each pricing location on the system.

In the day-ahead market, incremental offers and decremental bids (virtual supply offers and demand bids) can also be submitted, which indicate prices at which supply or demand are willing to increase or decrease their injection or withdrawal on the system. These INCs and DECs are tools market participants can use to hedge their positions in the day-ahead market.

From the offers and bids, the RTO constructs aggregate supply and demand curves for each location. The intersection of these curves identifies the market-clearing price at each location for every hour. Supply offers below and demand bids above the identified price are said to clear, meaning they are scheduled. Offers and bids that clear are entered into a pricing software system along with binding transmission constraints to produce the locational marginal prices (LMPs) for all locations.

Generators and offers scheduled in the day-ahead settlement are paid the day-ahead LMP for the megawatts accepted. Scheduled suppliers must produce the committed quantity during real-time or buy power from the real-time marketplace to replace what was not produced.

Likewise, wholesale buyers of electricity and virtual demand whose bids to buy clear in the day-ahead market settlement pay for and lock in their right to consume the cleared quantity at the day-ahead LMP. Electricity use in real-time that exceeds the day-ahead purchase is paid for at the real-time LMP.

Hour-Ahead Market

Not Applicable for PJM.

Real-Time Market

PJM must coordinate the dispatch of generation and demand resources to meet the instantaneous demand for electricity. Supply or demand for the operating day can change for a variety of reasons, including unforeseen generator or transmission outages, transmission constraints or changes from the expected demand. While the day-ahead energy market produces the schedule and financial terms of energy production and use for the operating day, a number of factors can change that schedule. Thus, PJM operates a spot market for energy, called the real-time energy market, to meet energy needs within each hour of the current day.

PJM clears the real-time energy market using supply offers, real-time load and offers and bids to sell or buy energy over the external interfaces. Real-time LMPs are calculated at five-minute intervals based on actual grid operating conditions as calculated in PJM's market systems. Generators that are available but not selected in the day-ahead scheduling may alter their bids for use in the real-time energy market during the generation rebidding period from 4 p.m. to 6 p.m.; otherwise, their original day-ahead market bids remain in effect for the real-time energy market.

Ancillary and Other Services

Ancillary services are those functions performed by electric generating, transmission and system-control equipment to support the transmission of electric power from generating resources to load while maintaining the reliability of the transmission system. RTO/ISOs procure or direct the supply of ancillary services.

PJM operates the following markets for ancillary services:

- Regulation: corrects for short-term changes in electricity use that might affect the stability of the power system.
- Synchronized reserves: supplies electricity if the grid has an unexpected need for more power on short notice.
- Day-ahead scheduling reserves (DASR): allows PJM to schedule sufficient generation to preserve reliability during unanticipated system conditions throughout the operating day.

Regulation service matches generation with very short-term changes in load by moving the output of selected resources up and down via an automatic control signal. In addition, PJM schedules operating reserves in the day-ahead market, and resources that provide this service are credited based on their offer prices. Reserve consists of 10-minute and 30-minute products.

Synchronized reserves are the equivalent of what is commonly referred to as spinning reserves, providing 10-minute reserves from a generator that is synchronized to the grid.

The DASR is the primary market mechanism for procuring the 30-minute reserves. A resource will only be assigned as amount of DASR corresponding to that amount of energy it could provide within 30 minutes of a request. If the DASR market does not result in procuring adequate scheduling reserves, PJM is required to schedule additional operating reserves.

Futhermore, two ancillary services are provided on a cost basis: (1) blackstart service, which helps ensure the reliable restoration of the grid following a blackout; and (2) reactive power, which supports the voltages that must be controlled for system reliability, are provided at cost.

Capacity Markets

Capacity markets provide a means for load-serving entities (LSEs) to procure capacity needed to meet forecast load and to allow generators to recover a portion of their fixed costs. They also provide economic incentives to attract investment in new and existing supply-side and demand-side capacity resources in PJM as needed to maintain bulk power system reliability.

PJM's capacity market is called the reliability pricing model (RPM). The RPM market was implemented in 2007 and is designed to ensure the future availability of capacity resources, including demand-resources and energy-efficiency resources that will be needed to keep the regional power grid operating reliability. RPM market design is based on three-year, forward-looking annual obligations for locational capacity under which supply offers are cleared against a downward sloping demand curve, called the variable resource requirement (VRR) curve. The VRR curve establishes the amount of capacity that PJM requires its LSE customers to purchase, and the price for that capacity, in each capacity zone (locational delivery area). Under RPM, when a locational delivery area is transmission-constrained in the auction (i.e., limited in the amount of generation that can be imported into those areas), capacity prices generally rise in that area relative to the overall PJM footprint.

Annual auctions are referred to as base residual auctions (BRAs). LSEs that are able to fully supply their own capacity need can choose not to participate in the auctions. Most capacity is procured through self-supply and contracted (bilateral) resources and the auctions procure any remaining needed capacity. To mitigate the exercise of market power, the RPM market rules provide a test to determine whather each capacity seller has market power. If the seller fails that test, that seller's bid id capped so as to replicate that seller's avoidable or opportunity costs.

Market Power Mitigation

In electric power market, mainly because of the largely nonstorable nature of electricy and the existence of tranmission constraints that can limit the availability of multiple suppliers to discipline markets, some sellers have the ability to raise market prices. Market power mitigation is a market design mechanism to ensure competitive offers even when competitive conditions are not present.

Market power may need to be mitigated on a systemwide basis or on a local basis where the exercise of market power may be a concern for a local area. For example, when a transmission constraint creates the potential for local market power, the RTO may apply a set of behavioral and market outcome tests to determine if the local market is competitive and if generator offers should be adjusted to approximate price levels that would be seen in a competitive market – close to short-run marginal costs.

The structural test for implementing offer capping in PJM is called the three pivotal supplier test. Generation is subject to offer caps when transmission constraints occur such that generators are run out of merit order, which means that a higher-priced generator must be run due to a transmission constraint that prevents the use of available lower-priced generation. When units are dispatched out of merit, PJM imposes offer capping for any hour in which there are three or fewer generation suppliers available for redispatch that are jointly pivotal, meaning they have the ability to increase the market price above the competitive level.

Price Caps

PJM has a $1,000/MWh offer cap in the energy markets.

Financial Transmission Rights

Financial transmission rights (FTRs) give market participants an offset or hedge against transmission congestion costs in the day-ahead market. An FTR is a financial contract protecting the holder from costs arising from transmission congestion over a path or a source-and-sink pair of locations on the grid. An FTR provides the holder with revenue, or charges, equal to the difference in congestion prices in the day-ahead market across the specific FTR transmission path. FTRs were originally formulated to protect LSEs from price uncertainty while redistributing excess congestion charges due to constrained conditions. Other market participants such as financial-only participants may purchase FTRs through the RTO's auctions or through secondary market purchases.

A related product is an auction revenue right (ARR). ARRs provide the holders with an upfront portion of the money raised in the FTR auctions. In general, they are allocated based on historical load served and can be converted to FTRs. As with FTRs, ARRs, too, give transmission owners (and eligible transmission service customers) an offset or hedge against transmission congestion costs in the day-ahead market.

Virtual Transactions

A virtual transaction allows a participant to buy or sell power in the day-ahead market without requiring physical generation or load. Cleared virtual supply (increment or virtual offers, or INCs) in the day-ahead energy market at a particular location in a certain hour creates a financial obligation for the participant to buy back the bid quantity in the real-time market at that location in that hour. Cleared virtual demand (decrement or virtual bids, or DECs) in the day-ahead market creates a financial obligation to sell the bid quantity in the real-time market. The financial outcome for a particular participant is determined by the difference between the hourly day-ahead and real-time LMPs at the location at which the offer or bid clears. Thus, through this financial arbitrage opportunity, virtual transactions in theory help to narrow the difference between the day-ahead and real-time prices.

Credit Requirements

Credit requirements balance the need for market liquidity against corresponding risk of default. Defaults within these markets are particularly troubling because losses due to default are borne by all market participants. PJM's tariff spells out the details for credit evaluations, credit limits, allowed forms of collateral and the consequences of violations or defaults.

To reduce financial risk, PJM's settlement cycle is seven days. The amount of unsecured credit allowed is $50 million for a member company and $150 million for an affiliated group. PJM does not allow unsecured credit in the FTR market.

Settlements

RTOs must invoice market participants for their involvement in their markets, including the amounts owed for buying and selling energy, capacity and ancillary services, and for paying administrative charges.

PJM has a two-settlement system, one each for the day-ahead and real-time energy markets.

SPP - Southwest Power Pool

Market Profile

Geographic Scope

The Southwest Power Pool Inc. (SPP) began operating in its real-time energy imbalance service (EIS) market on Feb. 1, 2007. Based in Little Rock, Ark., SPP manages transmission in nine states: Arkansas, Kansas, Louisiana, Mississippi, Missouri, Nebraska, New Mexico, Oklahoma and Texas.

In addition to operating its EIS market and managing open-access transmission facilities, SPP is the reliability coordinator for the NERC regional entity. As such, SPP enforces NERC-approved reliability standards for users, owners and operators of the bulk power system; coordinates reliability within and with neighboring areas; and ensures adequate reserves are procured within the SPP region. The reliability area is larger than the market area.

SPP's EIS market footprint includes 16 balancing authorities. Its members include investor-owned utilities, municipal systems, generation and transmission cooperatives, state authorities, independent power producers, power marketers and independent transmission companies.

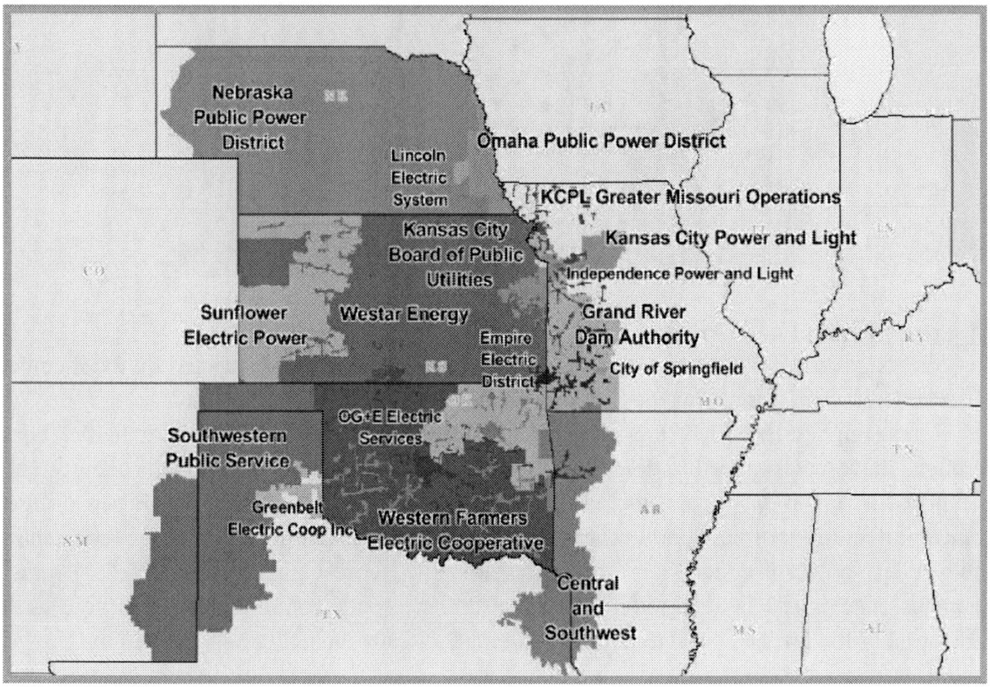

Southwest Power Pool (SPP).

Demand

SPP's record peak demand of 48 GW occurred in August 2011.

Import and Exports

SPP has alternate-current (AC) interties with the Midwest ISO, PJM Interconnection, Tennessee Valley Authority and Entergy Inc., among other systems. Additionally, SPP has two direct-current (DC) interties with ERCOT and seven DC interties to the western interconnect through New Mexico, Kansas and Nebraska. At times, SPP is a net importer of electricity and, at other times, SPP is a net exporter of electricity.

Market Participants

SPP's market participants include cooperatives, independent power producers, investor-owned utilities, power marketers, municipals, state agencies and transmission owners. SPP considers a participant in the EIS market to be an entity that has a legal and financial obligation to SPP in the market. Market participants must have generation assets to participate in the EIS market, or must directly represent an asset owner. Asset owners include generation companies and load-serving entities.

Membership and Governance

SPP is governed by a seven-member board of directors, with six elected by the members to serve three-year terms, plus the SPP president, who is elected by the board.

Supporting the board is the members committee, which provides input to the board through straw votes on all actions pending before the board. The members committee is composed of up to 15 people, including four representatives from investor-owned utilities; four representatives of cooperatives; two representing municipal members; three representing independent power producers and marketers; and two representing state and federal power agencies. The board is required to consider the members committee's straw vote as an indication of the level of consensus among members in advance of taking any actions.

Transmission

Owners

SPP transmission owners (TOs) are investor-owned utilities, municipals, cooperatives, state agencies and independent transmission companies. Some of the larger balancing authorities by installed capacity include:

- Southwestern Electric Power Co. (AEP West)
- OGElE Electric Services
- Westar Energy Inc.
- Southwestern Public Service Co. (Xcel Energy)
- Kansas City Power El Light Co. (Great Plains Energy)
- Omaha Public Power District
- Nebraska Public Power District
- KCPElL Greater Missouri Operations (Great Plains Energy)
- Empire District Electric Co.
- Western Farmers Electric Cooperative

Supply Resources

Generating Mix

By plant capacity, the generating mix includes these sources:

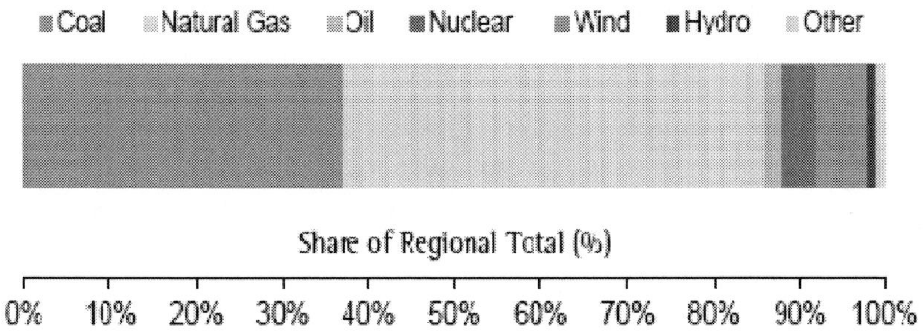

Market Features and Functions

Energy Markets

Day-Ahead

Not applicable.

Hour-Ahead

Not applicable.

Real-Time

SPP must coordinate the dispatch of generation and demand resources to meet the instantaneous demand for electricity.

SPP's EIS market provides market participants the opportunity to buy and sell wholesale electricity in real time.

If a utility requires more energy than it scheduled, the market provides the utility another option to buy the additional energy at real-time prices to make up the difference and meet its demand. Thus, participants use the EIS market to obtain energy available and offered from other utilities. Entities wishing to provide energy will submit offers to the market.

SPP uses a security constrained economic dispatch (SCED) to determine the lowest cost increment of energy that can be delivered to each location, considering the submitted offers, transmission limitations and system topology.

EIS market dispatch instructions are calculated for dispatchable resources, and locational imbalance prices (LIPs) are calculated for each settlement location (generation resource or load) on the system.

Resources are settled based on the LIP associated with their settlement location. Resources are only settled nodally. Load may choose to be settled either zonally or nodally. The LIPs are based on the resource offers and are locational.

Must-Offer Requirements

Not applicable.

Ancillary and Other Services

Ancillary services are those functions performed by electric generating, transmission and system-control equipment to support the transmission of electric power from generating resources to load while maintaining the reliability of the transmission system.

SPP does not offer an ancillary service market. However, it does require each transmission owner to provide or arrange for all of these services.

Capacity Markets

SPP does not offer a capacity market. However, it requires each market participant to have sufficient energy supply (capacity) to cover its energy obligations. SPP performs a supply adequacy analysis for each market participant based on a load forecast, resource plan, ancillary service plan and schedules received from market participants.

This analysis is performed for each hour of the next operating day, with results available by 3 p.m. of the day prior to the operating day.

Market Power Mitigation

In electric power markets, mainly because of the largely nonstorable nature of electricity and the existence of transmission constraints that limit the availability of multiple suppliers to discipline market prices, some sellers have the ability to raise market prices.

Market power mitigation is a market design mechanism to ensure competitive offers even when competitive conditions are not present.

Market power may need to be mitigated systemwide basis or locally.

Price Caps

Offers by market participants may not exceed $1,000/MWhour. This limit remains in effect until such time as SPP demonstrates in a filing with the Commission that sufficient demand response exists in the EIS market to allow a higher offer curve price limit or removal of the safety-net price limit.

Local Market Power Mitigation

When any transmission constraint is binding in the EIS market, SPP will screen the offer curve associated with resources on the importing side of each constraint. If the resource's offer is greater than the offer cap, then SPP will substitute the resource's offer with its offer cap.

Additionally, there is no mitigation for physical withholding in the EIS market, as the market is voluntary. The market monitor monitors determines whether the decisions to participate in the EIS market have a significant adverse impact on market outcomes.

FTRs

Not applicable.

Virtual Transactions

Not applicable.

Credit Requirements

RTOs must balance the need for market liquidity against corresponding risk of default. Defaults within these markets are troubling because losses due to default are borne by all market participants. SPP's tariff specifies credit rules needed to participate in the markets. These requirements provide for credit evaluations, credit limits, allowed forms of collateral and the consequences of violations or defaults.

Settlements

RTOs must invoice market participants for their involvement in their markets. Settlement is the process by which the RTO determines the amounts owed associated with buying and selling energy, capacity and ancillary services, and paying administrative charges.

The SPP settlement process calculates the quantity of energy imbalance for each asset (generation resource or load), calculates invoice dollars for energy imbalances and allocates over-and under-collection of revenues to asset owners. Settlement statements are published for each operating day. The market is facilitated so that SPP remains revenue neutral.

4. Financial Markets et Trading

Introduction

Financial markets affect physical natural gas and electric markets in key ways. In the past decade, the commodity markets associated with natural gas and electricity expanded dramatically, both in terms of volumes traded and the types of products offered. One result from this expansion has been to revamp the traditional relationship between physical and financial markets. The traditional view was that physical markets affect financial; financial products derive their value from physical products. Today, this is no longer the case. Physical markets continue to affect financial markets, but now, financial markets can affect physical markets – including prices – as well.

This chapter explores natural gas and electric commodity and capital markets.

Markets and Mechanisms

Financial markets are amorphous. These markets are not physical locations like grocery stores, or sites on the Web such as Amazon.com, where one can go to experience the financial marketplace. Instead, they are an array of products, mechanisms and participants that together flesh out the marketplace.

As mentioned, financial markets differ from physical markets in that no physical delivery occurs. This does not mean financial markets contain only investors and speculators; physical market participants enter the financial market to hedge. Similarly, it does not mean that financial markets involve only contracts that contain financial payout instead of physical delivery. Financial traders may use longer term physical contracts, but in a way that ensures no delivery will be required. Physical and financial markets are often closely intertwined and use the same market mechanisms.

Consequently, the best way to understand financial markets is to look at the market participants, products, market mechanisms and trading that together constitute the market.

Market Mechanisms

Transactions in both physical and financial markets are conducted through exchanges or over-the-counter (OTC). In the power market, trades may also be conducted in regional transmission organizations (RTOs, addressed in Chapter 3).

Exchanges

Exchanges are standardized. Trading on exchanges is subject to the rules of the exchange as well as law and regulation. The products traded are established in advance by the exchange, with no room for modification. The contracts specify pre-set locations, product types, quantities and trading mechanisms. Exchange rules typically permit bidirectional trading, or the ability to buy or sell with equal ease.

Trading in exchanges is conducted through electronic platforms, websites on which traders can buy and sell, or through trading pits where traders actively call out orders to buy and sell, known as open outcry.

On exchanges, buyers are not matched to specific sellers. Instead, exchanges are multilateral; buyers' needs may be met by a number of sellers, but the buyers and sellers do not interact or transact with each other.

Exchanges for gas and electricity are commodity exchanges, such as the New York Mercantile Exchange (Nymex), the world's largest physical commodity futures exchange. Nymex handles physical and financial natural gas products as well as financial power contracts.

Margin is the ability to trade without having to pay cash for the full value of the trade. Effectively, someone who trades on margin borrows much of the money used to buy or sell from the exchange or brokerage house. The trader posts collateral by putting down a certain amount of money or percentage of the trade value in cash or other items of value acceptable to the exchange.

Over-the-Counter (OTC) Markets

OTC markets are any markets that are not exchanges or RTOs. Transactions range from complicated individual negotiations for one-off structured contracts to standardized products traded through an electronic brokerage platform. The ability to tailor a contract to the exact needs of the counterparties is one of the chief benefits of OTC contracts.

OTC power transactions can occur in either traditional or RTO power markets.

OTC transactions are conducted through direct negotiations between parties or through brokers. Brokers range from voice brokers to electronic brokerage platforms such as that offered by the IntercontinentalExchange (ICE). Unlike an exchange, an electronic brokerage platform matches specific buyers and sellers, and is not anonymous.

Products may be negotiated individually or may be standardized. Many negotiations start with a standardized contract, such as that developed by the Edison Electric Institute (EEI) for power trading, and then modify it. Others start from scratch. Individually negotiated deals are called structured contracts.

To be tradable, a contract must include terms and conditions that make it attractive to more than one entity. Consequently, complicated, one-off contracts negotiated to meet the need of an individual seller and buyer may have little or no resale value. Typically, the standardized contracts traded on exchanges or electronic brokerage platforms are designed to be of interest to many market participants.

In general, when referring to transactions for physical and financial natural gas and power products, bilateral and over-the-counter mean the same thing.

The IntercontinentalExchange Inc. (ICE) operates as a global, electronic marketplace for trading both futures and OTC energy contracts. ICE's markets offer access to a range of contracts based on crude oil and refined products, natural gas, power and emissions, as well as soft commodities including cocoa, coffee, cotton, ethanol, orange juice, wood pulp and sugar, in addition to currency and index futures and options.

Regional Transmission Organizations

Electricity is also bought and sold through RTOs. In general, RTOs operate their markets to support the physical operation of the electric grid under their control, including making decisions about what generation to dispatch to meet customer demand. RTO markets are multilateral; buyers and sellers are not matched individually against each other. RTOs allow for bilateral physical transactions, although each RTO handles these differently. RTOs provide settlement, although this differs from the settlement offered by exchanges.

Also, RTOs use the word *clearing* to refer to the matching of supply and demand – to clear the market means the RTO accepts sufficient generation offers to meet demand. If a generator's offer in the day-ahead market clears, it means that generation was offered at or below the market clearing price and was chosen to generate the next day. RTOs do not take title to contracts and do not step into the middle of transactions, although they do maintain credit policies and allocate the costs of defaults or other performance failures across market participants.

RTO markets may have financial elements. One of these is virtual transactions. For example, a trader may offer generation in the day-ahead market, and the generation does not show up in the real time market. As a result, the trader is paid for his generation offer in the day-ahead market, based on day-ahead prices, but effectively has to pay to replace his power in the real-time market, paying the real-time price. Financial participants can be involved in virtuals; they use the physical product (generation offers or demand bids) in a way that results in no physical delivery. Virtuals are financial contracts, directly integrated into the RTO's operation of its physical market; they affect physical supply and demand, and prices. RTOs use them to add liquidity in an attempt to improve the efficiency of their physical markets.

RTOs may also offer financial transmission rights (FTRs) programs. Typically, a transmission owner that turns over operation of its transmission to the RTO wants certainty over its ability to flow electricity and about the cost of transmission. Because the RTOs operate using markets, this certainty cannot be provided directly. FTRs and similar instruments are designed to provide some degree of financial certainty to these transmission owners and firm rights holders. FTRs are linked to the physical operation of the RTO's system; transmission owners would be more reluctant to turn their transmission over to the RTO if they were going to experience significant costs due to transmission congestion resulting from other RTO members.

FTRs compensate the transmission owners or firm rights holders in a couple of ways. First, the RTOs auction off additional FTRs to others in the market, including financial participants who have no interest in buying, selling or transmitting physical power. The proceeds of the auction are returned to some of transmission owners or firm rights holders. The auction also determines a value for any FTRs held by the transmission owners or firm rights holders. FTRs can be bought and sold; the auction price gives an indication of their value for price discovery.

Other Market Mechanisms and Concepts

Leverage is the use of a small position to control or benefit a larger position. It increases the potential return, but also increases risk. Leverage can occur when a trader uses margin to trade.

Leverage can be used in other ways. As discussed in Chapter 5, some traders may try to use leverage to manipulate the market. For example, traders may use a smaller position in the physical market to benefit a larger position in the financial market. They may buy a financial product whose price is derived from a physical product. Then, they may try to buy or sell or otherwise influence the price of the physical product. If they succeed, their financial position benefits.

Liquidity refers to the trading and volumes occurring in a market. A market is said to be liquid if trading and volumes are such that any trader can liquidate his position at any time, and do so without affecting the prices. A market is thin if it has little trading or volume; in these instances, trading may affect prices. The benefits of liquidity are often used to justify practices that increase trading or volumes. However, not all trading or volumes are uniformly beneficial to markets and other market dynamics need to be taken into account.

Markets may not be uniformly liquid. For example, the market for the Nymex natural gas futures contract (NG) is generally thought to be liquid. However, when the United States Natural Gas Fund (UNG) became extremely large, its monthly process of getting out of the current contract and into the next involved selling and buying an extremely large volume of contracts at one time. If these transactions affected prices, then the market was affected.

Open interest is the total number of futures contracts in a delivery month or market that has been entered into and not yet liquidated by an offsetting transaction or fulfilled by delivery. Thus, as the clock is winding down during the settlement period, the open interest contracts (both in terms of the total number of contracts and the number of counterparties) are rapidly decreasing, so that a given number of contracts will represent an increasing share of the outstanding prompt-month contracts.

Clearing is a process in which financial or physical transactions are brought to a single entity, the clearing house, which steps into the middle of the transaction and becomes the counterparty to each buyer and seller. The clearing house assumes the risk that either the buyer or seller will fail to perform its obligations. Generally, clearing is used to manage counterparty risk. Clearing houses maintain rules about the creditworthiness of traders, collateral that must be posted and, of course, fees that must be paid for the service.

Settlement occurs at the end of a trading period, when the contract expires. At this time, delivery is to be made for a physical contract (physically settled) or a financial payout made for a financial contract (financially settled). Settlement occurs both in exchanges and in OTC trades. In OTC transactions, settlement occurs under the terms agreed upon by the parties. On exchanges, settlement occurs in a documented process and timeframe established by the exchange.

For example, every day at the close of Nymex trading, the NG contracts for forward months settle. Also, three business days prior to the start of the month of delivery (the prompt month), the contract expires and the last-day settlement (LD settlement) occurs in the last half-hour of trading. LD settlement is the final price for that particular futures contract term. Thus, the final settlement of natural gas futures prices on Nymex occurs on the third business day before the end of the calendar month in the last half-hour.

Most market participants avoid trading during the settlement period. As the time to termination approaches, price risk and volatility may increase, while market liquidity and the remaining open positions (open interest) are decreasing. For the Nymex natural gas futures contract (NG), most market participants either liquidate or roll their open long or short positions well before the settlement period. Rolling is the process of liquidating the current month's contract before it expires and purchasing a comparable position in the upcoming month. The trader holds the same number of contracts, but the contract month held changes as time passes and contracts expire.

Daily settlement prices are used to revalue traders' position to the current market price, for accounting and for margin calculations. Daily and LD settlement prices are also reported in publications and indexes, and are used for price discovery.

Mark-to-market (MTM) provides that at the end of each trading day, all trading positions are revalued to current market prices. This results in financial and accounting gains and losses. Traders can remove money resulting from gains from their accounts or use them for further trading; they do not have to liquidate their positions to get the money. Losses reduce the value of a trader's position, and may reduce the amount of collateral the trader needs to be able to trade on margin. If so, this may result in a margin call from the exchange or broker.

Mark-to-market is also an accounting transaction, in that a company's or trader's accounts are revalued daily to reflect changes in asset price. Losses can reduce the book value of a company or trader, and can affect its creditworthiness.

Trading is the buying and selling of contracts. Trades and transactions are virtually synonymous. Both refer to the buying and selling of power or natural gas.

Short selling is the selling of contracts a trader does not own, on the assumption that the trader will buy offsetting contracts prior to the contracts' expiration. This can be done on an exchange or other market that allows for bidirectional trading. Short selling has been of concern for potential market manipulation – traders sell a contract to drive the price down, and then buy when the price is low. Short selling is one of the ways market participants can trade futures financially – they sell the future, then buy it before the contract expires so the contracts net out and the trader faces no delivery obligation.

A **position** is the net holdings of a participant in a market. A trader's position in a specific instrument is combined purchases and sales of that contract. A trader's overall position is the combination of all positions in all contracts the trader owns. A trader's position is often referred to as the trader's commitment in the market.

Liquidating a position is the process of getting rid of a position. A trader who owns a contract will sell it to liquidate it. A trader who has sold a contract short will buy a contract to liquidate it. After liquidation, the trader holds no contracts.

Position limits have been imposed by ICE and Nymex, and are being developed for certain contracts by the CFTC. The position limits may restrict the number of shares a trader may hold in a particular investment at any point in time, during the month the contract expires, or during some period closer to settlement. For example, CME imposes accountability levels for any one month and for all months, and has limits for expiration-month positions. CME's limits for the Nymex NG contract are12,000 contracts for all months, 6,000 contracts for one month and 1,000 contracts in the expiration month. Trading entities can petition to have these waived or modified.

Volumes give an indication as to the nature of the activity occurring in the market at any point in time. Volume can be expressed in a number of ways. It can be the number of transactions executed during a specified period of time or the volume of the product contained in the contracts.

Volumes give market participants information about what is going on in the market. For example, if many Nymex contracts are traded, and the number of trades is relatively few, market participants know that a relatively few traders are making high-volume trades. Conversely, if a high number of Nymex contracts are traded and the number of trades is also high, then at least some of the trading is being done in small volumes. This could result from broad interest in the market – lots of active traders – or it can result from relatively few traders making a lot of trades.

Market Participants

Financial markets are used by many types of participants. These markets represent an opportunity for physical players, producers and marketers to buy or sell some physical products or to hedge physical supplies and obligations with physical or financial products. Investors, speculators and investment funds also use these physical and financial products for financial gain.

Products

Products, for purposes of trading, are contracts — also known as securities or instruments — that can be bought and sold. Contracts for physical trading in natural gas or electric markets provide for the delivery of natural gas or electricity. The actual molecules of gas or electrons may be delivered as a result of the contract. Financial contracts do not provide for delivery of a product; instead, they provide a financial payout.

Consequently, what traders buy and sell are contracts giving them a right or obligation. For physical contracts, this is the obligation to deliver or take delivery of natural gas or electricity in exchange for payment. For financial contracts, it is the right to a payout in exchange for payment.

Other physical and financial products give traders rights to buy or sell a contract in the future at a given price – an option to buy or sell.

The word *derivative* is used for a category of contracts whose value is derived from some other physical or financial product or contract. Standardized derivative contracts trade on exchanges such as Nymex. Financial contracts are derivatives. A common financial derivative used in natural gas and electric markets is the swap.

The CFTC also considers futures contracts to be derivatives. As futures contracts approach expiration, their price should begin to mirror spot prices – to derive their price from spot prices. However, at other times, futures contracts are simply the price parties are willing to pay for natural gas at some point in the future and do not derive their value from any other product or contract. In fact, expiring futures contracts may affect spot prices, as well.

Instrument Basics

Each instrument is traded in its own market and is identified by the market name, such as spot or futures. Each market and instrument has characteristics such as time-frame, location, contract type, product conveyed by the contract and, for swaps, the mechanism for determining the payout.

Product conveyed: Each contract specifies what it is that is being bought and sold. For physical contracts, this would be natural gas or electricity. For derivatives, it may be a payout derived from natural gas or electricity. All contracts conveying or derived from natural gas, for example, would be in natural gas markets.

Time: Each contract has a number of time elements.

The trade date is the date on which the contract is written (typically the date the trade is executed).

The expiration day is the last day for a contract, after which it is no longer available to be bought and sold; it is often the same day as the settlement day. Exchanges and electronic brokerage platforms may also impose a termination date, the last date on which a contract may be traded.

Physical contracts also specify the delivery day(s) or month – the day(s) or month during which the product is to be delivered.

For physical products, begin and end dates are the dates for which a physical product (natural gas or electricity) is to be delivered. For financial products, these dates address the contracts whose prices are used to set the payout. For example, a next-day physical gas deal may have a trade date of Aug. 7, a begin date of Aug. 8 and an end date of Aug. 8. A monthly product may trade on Aug. 7, its trade date; the flow of natural gas would have a begin date of Sept. 1 and an end date of Sept. 30.

For the Nymex NG contract, the termination day and settlement day are the third-to-last business day of the month before the month in which the gas is to be delivered. The settlement period occurs from 2 p.m. to 2:30 p.m. on the termination day.

Short-term or spot contracts provide for delivery or payout during the current or next day; the price for these contracts is known as the spot price.

Daily physical contracts are for delivery on a given day or set of days.

Electric physical and financial contracts may also specify peak or off-peak delivery, with the peak or off-peak hours defined by the contract.

Contracts for delivery a month or more into the future are forward contracts, or if they are traded on exchanges, futures contracts. The Nymex NG contract, for example, provides for the delivery of 10,000 MMBtu of natural gas in the month specified by the contract. Contracts are offered for every month over the next 12 years.

Monthly contracts are referred to by how close they are to expiring. Spot month is the current month. Prompt month is the month after the spot month or current month – it is the next trading month. For trading in January, February is the prompt month.

Another time element is the delivery or payout period, such as daily, next day or monthly. Monthly contracts generally are for delivery in equal parts over a month at a specified price for gas and for the contracted amount in each hour for power.

Location: All physical contracts specify the location where the natural gas is to be delivered, such as the Henry Hub in Louisiana. Financial contracts also have a locational element, determined by the underlier. For example, if a financial derivative uses the Nymex gas contract as its underlier, the derivative's locational element is the Henry Hub.

For natural gas, the locations are referred to as market hubs, which are located at the intersection of major pipeline systems. For power, contracts are often based on locations known as nodes, zones or hubs.

For gas, the principal hub and pricing point is the Henry Hub, which is used for all Nymex gas futures contracts and is the reference point for overall prices in the United States. Prices for other locations are often references as a difference from Henry Hub, known as basis.

Products traded on exchanges and preset products traded on OTC electronic brokerage platforms such as ICE use standardized locations or pricing points. Locations for other OTC transactions use whatever location the counter-parties to the contract desire. For physical contracts, the location must be physically viable.

For financial products, it can be whatever the parties desire (although complicated locations make pricing more difficult due to the lack of reference points for price discovery).

Quantity: All physical contracts specify the amount of natural gas or electricity to be delivered. For contracts traded on an exchange or for preset contracts traded on an OTC electronic brokerage platform, the quantity is predetermined and specified in the contract. For bilateral contracts traded in OTC markets, the quantity contained in the contract can be anything the parties want it to be. For standardized products traded on electronic brokerage platforms, the quantity is fixed.

Price: The price paid for a contract is usually that set by the market and is usually known at the time the contract is bought or sold.

Fixed prices are known at the time the transaction is entered into – it is the price at which the seller agrees to sell and the buyer agrees to buy. Contracts sold at fixed prices are typically paid for at the time of purchase.

Floating prices are set by formulas pegged to something whose price is not currently known but which will be at the time the contract expires, such as an index. For example, a price may be tied to the average of the all of the daily prices at a location over the course of a month, typically as published in an index. An index contract is a commonly traded instrument based on major trading points, such as the Houston Ship Channel or the Henry Hub.

Spot price is a cash market price for a physical commodity that is available for immediate (next day) delivery, and may be reported to publishers for indexes.

Standardized forward contracts and futures contracts are traded for every month, years into the future; the NG contract is traded 12 years into the future. Each of those contracts for which trading has occurs has a price. Together, the prices for future contract months creates a trajectory of prices known as forward or futures curves.

The settlement price is effectively the final official reported price for certain contracts and is an average of prices for trades occurring during the settlement period. The settlement price forms the basis for payout in financial derivatives that use the contract as its underlier, for margin calls and for reporting to index publishers and other entities.

For example, the natural gas futures contract settlement price is the weighted average price of all sales made during the contract's 30-minute settlement period – the last 30 minutes of trading on the contract's termination day. The gas futures final settlement price sets, in whole or in part, the payout for financial derivatives that use the contract as its underlier.

Physical Products

Physical products involve physical delivery. Power products include energy, transmission (firm and nonfirm) and ancillary services. Electric energy products include spot transactions, full requirements sales and bundled services, among other things. Natural gas products include the natural gas molecules themselves, transportation and storage.

Forward products are contracts for delivery in future months traded through the OTC market (including electronic brokerage platforms). If the product is traded on an exchange, it is known as a futures contract.

A futures contract is a standardized forward contract traded on a regulated exchange. Each contract represents the same quantity and quality of the underlying physical commodity, valued in the same pricing format, to be delivered and received at the same delivery location. In addition, the date of delivery and receipt is the same for all contracts traded for a particular calendar month. The only element of futures contract that is subject to change when it is bought or sold is the price.

For the natural gas industry, the dominant futures contract is the Nymex NG contract. For the NG contract, the standard contract specifications are the delivery location – Sabine Pipeline Hub at the Henry Hub in Louisiana; the term – monthly; and the quantity – 10,000 MMBtu delivered equally over the course of the month. Not all forward contracts have fixed prices. Some involve trades executed now to buy or sell at some point in the future, at a price to be set in the future. One example of this is a forward physical index contract. This OTC contract obligates one party to buy the underlying commodity or security and the other party to sell it, for a delivery price to be determined when a specific index sets at some known date in the future. Many natural gas purchases are made under forward physical index contracts; among other things, it may provide state regulators with some assurance that the price paid is reasonable. Forward and futures contracts with fixed prices can be used for price discovery, hedging or speculating. They may be traded by any of the participants listed earlier. Physical participants may use forwards or futures to obtain gas or electricity for delivery in some future month, or may use them to manage the risk of – or, hedge – their physical positions. Futures contracts that go to delivery lose their anonymity at settlement. However, only a small fraction – often, less than one percent – of futures contracts go to delivery.

Financial traders may buy or sell futures contracts for financial purposes, as exchanges have bidirectional trading – markets in which trader can buy and sell contracts with equal ease. Bidirectional trading allows the sale of contracts a trader does not own, known as short sales. A financial trade may buy a forward or future, then either sell the contract later or neutralize it by obtaining an offsetting contract. Because the two offset, the trader has no physical delivery obligation. Most Nymex natural gas futures contracts do not go to delivery.

Physical products can be combined to create different physical positions for use in physical and financial trading. A price spread can be created using forwards priced at indexes for two different hubs. The trader would buy physical natural gas at one index and sell at another. For example, the trader buys gas priced at the Houston Ship Channel index (and would have to take delivery of the gas there) and sells gas at the Texas East M-3 index (and would have to deliver it there). The trader earns the difference between the two contracts. A physical spread carries with it the obligation to make or take physical delivery of natural gas at both points, so pipeline capacity would be required to actually move gas between these points. A financial trader could also execute this trade, but would have to unwind both positions before delivery.

Indexes are formally published for both natural gas and power using a methodology posted by the publisher. An index may be used to set the price for settlement of floating price contracts. Indexes are also used by a variety of market participants to inform their decisions in the many steps in the electric or natural gas supply chain or in trading, known as price discovery. Data used in indexes are submitted voluntarily by firms involved in trading.

Indexes are commonly formed using volume-weighted average prices.

Financial Products

Financial contracts do not provide for the delivery of a product, but instead provide a financial payout. This is often based on the value of some physical or financial product specified by the contract, called the underlier. The value of these financial contracts is derived from the value of the physical or financial instrument specified in the contract as the basis for payout; as such, they are derivatives.

A key benefit of financial products is that they have no physical delivery and they are self-liquidating. Speculators who trade futures have to undo their position to eliminate the delivery obligation. One who trades derivatives, on the other hand, does not bear the complications of unwinding positions; he can simply wait for expiration and receive or pay the contract's payout.

Swaps

A key financial contract structure used in natural gas and electric markets is the swap, or contract for differences. The CFTC defines a swap as an "exchange of one asset or liability for a similar asset or liability ... it may entail buying ... on the spot market and simultaneously selling it forward. Swaps may also involve exchanging income flows..." Effectively, a swap is the exchange of like for like. Consequently, physical instruments cannot be swaps because parties to physical goods pay or receive money in exchange for delivery of the physical good. However, the exchange of money in terms of payment and payout constitutes a swap.

Options

An options contract conveys a right (but not the obligation) to buy or sell something else. It comes in two forms: the right to buy or the right to sell something at a specified price at or before a specified date. The buyer buys the right – the option – to buy or sell in the future; the seller (or writer) sells the obligation to sell or buy if the buyer exercises his right.

An option to buy is known as a call option; an option to sell is a put option. The price paid to buy or sell the option is known simply as the option's price. The price at which the option may be exercised is the strike price. Electing to buy or sell the underlying commodity or security is known as exercising the option.

Options traded on exchange or electronic trading platforms may be traded up to their expiration. Consequently, the owner of an option may sell it rather than exercising the option or letting it expire.

Traders buy and sell options for a number of reasons. First, they provide a risk management tool akin to insurance. Second, traders may use options traded on exchanges or electronic trading platforms to speculate. For example, a speculator may trade an option and hope to gain from price movements, akin to how they might trade other contracts, such as futures. If a trader buys an option, he can sell it up to expiration and pocket the difference between the purchase price and the sales price. Further, as in futures, the seller of an option traded on an exchange can offset his obligation by purchasing the offsetting option, thereby eliminating the risk of the contract going to delivery.

Finally, traders may use options to boost their trading income or to reduce the volatility of their returns. Options require less money up-front than a futures contract or swap, which can be a benefit to traders with limited funds.

Trading and Transacting

Trading Mechanics

Market prices are the collective result of individual trades. Open interest is the aggregation of traders' positions.

Trading is the buying and selling of contracts. A trade is a single purchase or sale. A position is the accumulated unexpired contracts purchased or sold, at a point in time. Traders may have positions in each contract, as well as an overall position reflecting all their contracts.

Trading requires a buyer and a seller, each willing to transact for a price. A buyer bids a price he is willing to pay to purchase a contract; this is the bid price. A seller offers his product for sale; the price at which he offers it is the offer price.

These prices may or may not be the same. When they differ, the distance between them is the bid-offer or bid-ask spread. This spread is the difference between the highest price at which buyers are currently willing to buy (the highest bid) versus the lowest price at which sellers are currently willing to sell (the lowest offer). For example, if a buyer bids $7 and the seller offers at $10, the bid-ask spread is $3.

Trading Concepts

Traders need to know how their trades and positions will be affected by market changes. One way this is done is by considering whether a trade or position benefits or loses when prices go up or down. A position is long if it benefits from increases in price. It is short if it benefits from falling prices. If it is neutral, benefitting from neither a rise nor a fall in prices, it is said to be flat.

For example, a trader who purchases an NG contract is going long; that contract will benefit from increases in price. A trader who sells the contract is going short; the trader will benefit from falling prices.

The concept of being long or short applies to other forms of transactions. Absent anything else, a generator is long electricity; a consumer short electricity. If the generator obtains a contract to sell electricity to the consumer at its cost of generating, the generator is flat.

The task of identifying long or short is not always easy. A trader may have a variety of positions in a number of contracts, some long and some short. How the overall position benefits from swings in prices depends on each of the components and how they interact with each other.

Trading Strategies

Traders decide what products to trade, how to trade them and in which combinations. Their strategies will depend on their objectives. Broadly, market participants trade to accomplish any of three objectives: to buy or sell physical products, such as natural gas or electricity; to manage the risk of their physical positions, or hedge; or to make money.

Hedging

Market participants who are in the market to buy and sell natural gas and electricity are interested in making money through their physical operations. These physical operations determine their individual risks and hedging needs.

Thus, each physical market participant's risks depend on his role in the physical delivery and consumption of natural gas and electricity. A natural gas producer has different risks and therefore different hedging objectives than an LDC that needs to purchase gas to resell to retail consumers.

An LDC, for example, is concerned with obtaining sufficient volumes to serve variable customer demand and in the price paid for those volumes. A producer may be concerned about selling all his output (unless he can store it), and about the price he will be paid for the gas. Physical market participants may have other concerns as well. Producers may need a predictable cash flow to support their financing. LDCs may be concerned with state regulators determining that their gas purchasing practices were imprudent.

Such concerns drive both procurement and sales decisions as well as risk management decisions. The two are often closely interconnected. For example, an LDC needs to buy enough gas to meet extremely variable retail demand, but not too much. Healso wants a price that regulators and consumers will see as reasonable. Consequently, this LDC develops a procurement and risk management – hedging – strategy taking these factors into account.

To purchase sufficient quantities, the LDC may create a portfolio of supplies, with a block of firm supply to meet minimum daily needs. Hemay also decide to buy in the spot market to meet demand peaks. Hemay diversify the sources of gas, both to improve reliability of supply but also to diversify its price.

He can also manage his risk financially. In the commodities and securities market, a hedge is a transaction entered into for the purpose of protecting the value of the commodity or security from adverse price movement by entering into an offsetting position in a related commodity or security. 1-ledging is used when describing the purpose of entering into a transaction with the intent of offsetting risk from another related transaction.

Speculation

Traders seeking to make money fall into a couple of categories: investors and speculators. These categories are distinguished by the strategies they use to profit from the market. Investors are relatively passive; they are in the market to benefit from long-term price movements and to diversify a broader portfolio. Speculators actively seek to gain from price movements.

Trading Analysis

In deciding whether to trade, both hedgers and speculators pay attention to what is going on in the market, and develop their own view of where the market is likely go. They may develop complicated forecasts as the basis for decisions on a number of transactions: whether, when and where to build a merchant power plant, how to hedge natural gas production, and of course, when to buy and sell in the markets.

Two general schools influence traders' thinking when analyzing markets for trading opportunities. The first is fundamental analysis, which takes into account physical demand and supply fundamentals including production, pipeline and transmission capacity, planned and unplanned outages, weather and economic and demographic changes. Changes in information about fundamentals (or changes in perceptions of fundamentals) alters traders' view of the supply-demand balance, and therefore, of prices. Fundamental analysis is used often to determine the impacts of longer term trends in the physical market – the development of shale gas supplies, for example.

The second school of thought is technical analysis, which forecasts price movements based on patterns of price changes, rates of change, changes in trading volumes and open interest, without regard to the underlying fundamental conditions. Instead of looking at the market for a physical good, technical analysis looks at trading and price changes. These quantitative methods have become a dominant part of market analysis. Technical analysis is used most often to determine short-term movements and trends, helping traders time their buys and sells.

Capital Markets

Capital markets provide the money to make investments in infrastructure such as power plants or natural gas pipelines, to operate plants and companies and to trade or conduct transactions. Access to capital depends both on the health of capital markets and also on the perceived riskiness of the entity seeking the capital. To measure relative riskiness, many providers of capital look at different measures, including credit ratings assigned by the three major crediting rating firms: Standard and Poor's (S&P), Moody's and Fitch.

Capital Expenditures

One effect capital markets have on energy markets is in capital spending – undertaking work or investments that require capital. The recent recession and shake-up in capital markets took a toll on capital spending as fi nancial commitments to infrastructure in 2009 fell for the fi rst time in years but then rose again (see bar chart).

The electric industry makes up the bulk of the capital expenditures expected by the energy companies, as compiled by SNL between 2010 and 2012 (see pie chart, page 123). These capital additions lie primarily in generation, transmission and distribution. In the past couple of years, investment decisions on new big-dollar projects have been delayed.

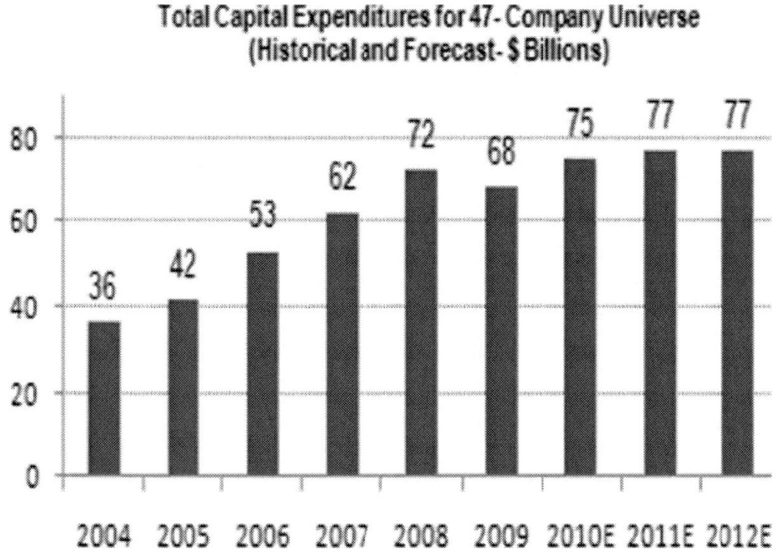

Source: SNL Energy.

Types of Capital

Capital comes from two general sources of financing – equity and debt.

Debt financing involves borrowing money to be repaid over time, along with interest at a fi xed or variable interest rate. With debt, the investor does not become an owner of the company. Some common types of debt include bonds – securities that companies issue in fi nancial markets with maturities (when the loan has to be repaid) of more than a year; shorter term debt issued by companies through financial markets; and bank loans, such as lines of credit. A revolving line of credit is an assurance from a bank or other institution that a company may borrow and repay funds up to some limit at any time. Municipal and cooperative utilities typically use debt; they have no ownership to sell.

Characteristics of debt include:

- Capital obtained through debt must be repaid or refinanced.
- Debt may be short-term, such as lines of credit from banks or corporate paper, or it may be long-term.
- Companies must make their interest payments and repayment on schedule, or the debt holders can take action, including forcing the company into bankruptcy. A company must generate sufficient cash through its operations or through other financing to make these payments.
- Interest gets paid before equity dividends.

- Interest payments are tax deductible.
- Debt gives lenders little or no control of the company (unless it gets into financial trouble).
- Debt can leverage company profits; similarly, it can magnify losses.
- Lenders are typically conservative, wanting to minimize downside risks.
- Borrowers may be required to pay collateral to secure debt. Debt without collateral is known as unsecured debt.

Equity financing is money provided in exchange for a share in the ownership of the business. A company does not have to repay the capital received, and shareholders are entitled to benefit from the company's operations, perhaps through dividends.

- Equity capital can be kept by the company indefinitely.
- Companies can issue shares in the company – stock – through financial markets. They may also use private equity – money from venture capital firms or private investment companies that buy into a company and which may or may not take an active role in operating the company.
- The most common form of stock is common stock, which does not require regular payments, but it may receive dividends; investor-owned utilities typically pay dividends.
- Equity does not provide a tax deduction to the company; dividends and other payouts are not tax deductible.
- Stockholders and private equity investors get a say in how the company is operated and may impose restrictions.
- Equity investors may be more willing to assume higher risks in return for a higher potential returns. Utilities are typically considered fairly conservative investments. Natural gas producers attract a more risk-inclined investor.
- The return required to attract equity is higher than the interest paid to debt holders.
- Equity capital does not require collateral; it gets a share in the company.
- Additional equity capital infusions may dilute, or reduce, the value of existing shares.

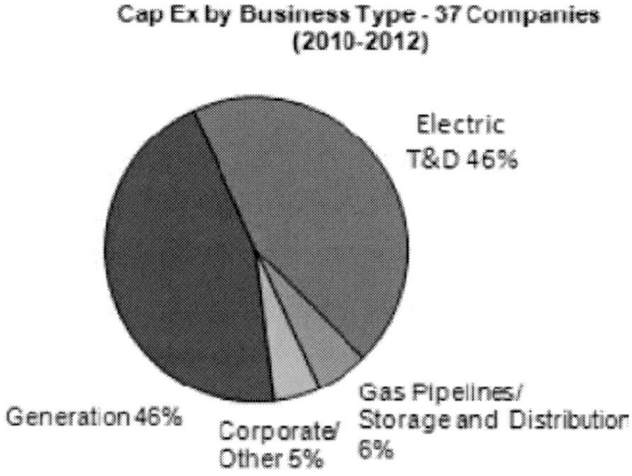

Source: SNL Energy.

Companies often try to match the type of financing with the investment they are making. Pipelines, power plants and transmission facilities are long-lived assets. They are typically financed using long-term capital, such as stock and long-term bonds, which can have 30-year maturities.

Other capital is needed to conduct day-to-day operations. Some of the cash needed to fund operations comes from a company's revenues. However, revenues do not always come in when payments are due.

Consequently, companies also rely on working capital. This can include some long-term capital from stocks and medium- and long-term bonds. Short term investments and day-to-day operations also rely on commercial paper and bank loans to cover day-to-day cash needs. If a company faces significant problems, it may have to issue especially high-priced debt – junk bonds – to obtain financing. These are bonds issued by entities lacking investment grade credit ratings (see below).

In the past few years, private equity investors and hedge funds – individuals or funds seeking to take ownership positions in companies or even buy companies – have taken an interest in the energy industry, and have provided another source of equity financing. Private equity firms have bought energy companies, including Puget Sound Energy, Energy Transfer Partners and Mountaineer Gas Co.

Two of the larger private equity firms interested in the energy sector include Kohlberg Kravis Roberts et Co. (KKR) and Macquarie Group.

Credit Ratings

Not all companies (or governments) present the same riskiness to investors. Investors, traders and others consider the risks their counterparty may present, including the risk of default. One standardized tool used to assess relative risk is the credit rating. Credit rating agencies, such as Standard and Poor's, Moody's and Fitch, assess a company's riski- ness every time it wants to issue bonds. A credit rating represents the likelihood that an issuer will default on its financial obligations and the capacity and willingness of a borrower to pay principal and interest in accordance with the terms of the obligations. Many organizations, including RTOs, consider bond ratings, among other things, when setting their credit policies, which determine with whom companies may transact and whether the counterparty will need to post collateral. Each credit rating agency has its own way of assessing risk, reflected in the rating system they use.

Ratings by Industry Sector

Electric utilities largely are rated investment grade, with ratings of BBB or better.

Merchant generators include generating companies that are completely unaffiliated with integrated utilities (and are known as independent power producers, or IPPs) and those that are affiliated but which receive at least half their cash flow from competitive power sales. The affiliated companies typically have higher ratings; SEIP views the integrated merchants' business profile scores as strong or satisfactory. SEIP typically rates IPPs fair or weak.

The midstream sector of the natural gas industry, which contains pipelines, processing plants and storage facilities, is also typically rated investment grade. Midstream companies' ratings average BBB and are said by rating agencies to have a stable outlook.

5. Market Manipulation

Energy markets can be manipulated. Following the energy crisis in the western United States early last decade, Congress granted the Federal Energy Regulatory Commission new authority to address this threat to the integrity of its regulated markets. At Congress's direction, the Commission enacted a catch-all antifraud rule that is modeled on the Securities and Exchange Commission's (SEC's) decades-old rule protecting the securities markets. Recognizing that other regulators have long prohibited manipulation of other markets such as securities and commodities, the Commission draws from the experience of sister federal agencies in implementing the Commission's anti-manipulation authority.

Manipulation comes in many varieties. As a federal court of appeals has stated in the context of commodities manipulation, "We think the test of manipulation must largely be a practical one.... The methods and techniques of manipulation are limited only by the ingenuity of man." The Commission recognized this reality by framing its Anti-Manipulation Rule broadly, rather than articulating specific conduct that would violate its rules. While manipulative techniques may be "limited only by the ingenuity of man," the following are broad categories of manipulations that have surfaced in the securities and commodities markets (including the energy markets) over the years. The borders of these categories are not clearly defined and some can belong to multiple categories, such as wash trading (i.e., buying and selling identical stocks or commodities at the same time and price, or without economic risk). Traders may also combine elements of various schemes to effect a manipulation.

Withholding

Withholding is the removal of supply from the market and is one of the oldest forms of commodities manipulation. The classic manipulation of a market corner involves taking a long contract position in a deliverable commodity and stockpiling physical supply to force those who have taken a short position to buy back those positions at an inflated price.

Withholding played an important role in the western power crisis that engulfed California in 2000. Market participants, particularly Enron, exploited supply-demand imbalances and poor market design. Generation operators scheduled maintenance outages during peak demand periods, which is an example of physical withholding. In addition, transmission lines were overscheduled to create the appearance of congestion in an effort to reduce the supply of electricity. The result of these efforts in combination with economic withholding and information-based schemes discussed below was that wholesale electricity prices soared. Utilities such as Pacific Gas El Electric (PGElE) and Southern California Edison were unable to pass on these high prices to their retail customers because of state price caps. The crisis led to widespread blackouts, heavy losses to the state's economy and the bankruptcy, in April 2001, of PGElE.

Economic withholding, which also contributed to the western power crisis, is similar to physical withholding, but rather than turning off a generator or stockpiling a physical commodity, the manipulator sets an offer price for a needed resource that is so high that the resource will not be selected in the market. For example, a generator in a constrained market such as New York City could purposely set its offer price high enough that it would not be called on to run. This scheme would create a shortage of generation and, thus, would raise prices for the benefit of the rest of its generation fleet or its financial positions. If done to benefit financial positions, this scheme would be a cross-product manipulation similar to the ones discussed later.

Information-Based Manipulations

Many manipulative schemes rely on spreading false information, which involves knowingly disseminating untrue information about an asset's value in order to move its price. A well-known scheme is the *pump and dump,* in which a participant spreads a rumor that drives the price up and then sells the shares after the price rises. In the energy markets, a common way to misrepresent a commodity's value is to misrepresent the price of the commodity or its level of trading activity.

False reporting and *wash-trading* schemes were well-documented information-based manipulations that took place in the early 2000s and contributed to the western energy crisis. False reporting occurs when a market participant submits fictitious transactions to a price-index publisher to affect the index settlement price.

Similarly, wash trading involves actual but offsetting trades for the same (possibly nonmarket) price and volume between the same market participants such that no economic exchange takes place; however, it falsely inflates trading volumes at a price level and gives the impression of greater trading activity.

False reporting and wash trading have resulted in a number of criminal prosecutions by the Department of Justice. A variation on these practices is *round-trip trading*, in which a trader sells an asset but agrees to buy it back at the same time.

Manipulative Trading Techniques

A number of manipulative trading techniques that have arisen in securities and commodities trading may be subject to the Commission's Anti-Manipulation Rule. Traders may seek to inflate trading volumes or trade at off-market prices to serve purposes such as maintaining market confidence in a company's securities or to move a security's price to trigger an option.

Marking the close is a manipulative practice in which a trader executes a number of transactions near the close of a day's or contract's trading to affect the closing or settlement price. This may be done to obtain mark-to-market marks for valuation, to avoid margin calls or to benefit other positions in related instruments, the latter of which was done by both Brian Hunter and Constellation Energy Commodities Group (discussed in the next section). *Banging the open* is a similar practice in which a trader buys or sells a large quantity at the opening of trading to induce others to trade at that price level and to signal information on fundamentals. Other manipulative trading techniques exist, and previously discussed practices like wash and round-trip trading fit under this description as well.

FERC Investigation and Prosecution of Cross-Product Manipulations

Manipulators have grown more sophisticated with the expanded use of derivative products, whose value is set by the price of transactions in a related product. Many of the manipulative schemes that staff has investigated and prosecuted are cross-product schemes in which an entity engages in price-making trades in the physical market, often at a loss, to affect the settlement price of price-taking derivative instruments.

Brian Hunter and Amaranth Advisors

In 2010, a FERC administrative law judge (ALJ) found that Brian Hunter, a trader with Amaranth Advisors LLC, manipulated the settlement price of the March, April and May 2006 New York Mercantile Exchange natural gas futures contracts (Nymex NG contract). Hunter bought large long positions in Nymex NG contracts, which he sold rapidly during the contract's final settlement period with the intent of pushing down the settlement price. He engaged in this behavior while he concurrently held larger short positions in financial look-alike contracts, principally on the IntercontinentalExchange (ICE), which benefited from a lower Nymex NG price. The Commission affirmed the ALJ's findings and imposed a $30 million civil penalty on Hunter. His appeal of this decision was pending at the time of this printing. Other Amaranth parties had previously settled.

Energy Transfer Partners

In 2007, the Commission issued an order to Energy Transfer Partners LP (ETP), directing it to show cause why the Commission should not find that ETP violated the Commission's market behavior rule that prohibited manipulative practices. The order charged that ETP manipulated the Houston Ship Channel (HSC) market for natural gas. FERC charged that ETP entered into financial and physical positions at HSC that profited from a lower HSC index price. ETP's financial positions were mostly basis swaps that exchanged the Henry Hub price for the HSC index price with a differential. ETP's physical positions were mostly purchases of natural gas at the HSC index price. The HSC index was set by physical fixed-price trades made during bidweek, which is trading during the last few days of one month for gas to be delivered every day of the following month. These trades were reported to Platts, a news and price-reporting service, and calculated according to Platts's methodology. The FERC order charged that ETP entered into bidweek physical trades at less than competitive prices with the intent to drive down the HSC index price. ETP's trading volumes were so great that it dominated the HSC index. On the eve of trial, ETP agreed to pay $30 million to settle the allegations.

Constellation Energy Commodities Group

In 2012, the Commission approved a settlement with Constellation Energy Commodities Group (CCG) in which CCG agreed to disgorge $110 million in unjust profits and pay a civil penalty of $135 million. FERC staff had alleged that CCG entered into significant loss-generating physical and virtual day-ahead transactions in electricity markets in and around New York state with the intent to move day-ahead price settlements to benefit financial swap positions that received their prices from those settlements.

Details of FERC Jurisdiction

The Energy Policy Act of 2005 added anti-manipulation provisions to the Federal Power Act, 16 U.S.C. § 824v (2006), and the Natural Gas Act, 15 U.S.C. § 717c-1 (2006), which the Commission implemented in Order No. 670. This rule has been codified as 18 C.F.R. § 1c (2011) (Anti-Manipulation Rule).

The Anti-Manipulation Rule prohibits anyone from (1) using a fraudulent device, scheme or artifice, or making any untrue statement of a material fact or omitting to state a material fact necessary to make a statement that was made not misleading, or engaging in any act, practice or course of business that operates or would operate as a fraud or deceit upon any entity; (2) with the requisite scienter (that is, an intentional or reckless state of mind) (3) in connection with a transaction subject to FERC jurisdiction. The Commission need not show reliance, loss causation or damages to prove a violation. The Anti-Manipulation Rule applies to any person, entity or form of organization, regardless of its legal status, function or activities.

The prohibition is intended to deter or punish fraud in wholesale energy markets.

The Commission defines fraud in general terms, meaning that fraud includes any action, transaction or conspiracy for the purpose of impairing, obstructing or defeating a well-functioning market. Fraud is a question of fact that is to be determined by all the circumstances of a case. In Order No. 670, the Commission found it appropriate to model its Anti-Manipulation Rule on Securities and Exchange Commission (SEC) Rule 10b-5 in an effort to prevent (and where appropriate, remedy) fraud and manipulation affecting the markets the Commission is entrusted to protect, while providing a level of certainty to market participants that is beyond that which the Commission would be otherwise required to provide. Like SEC Rule 10b-5, FERC's Anti-Manipulation Rule is intended to be a broad antifraud catch-all clause.

The Commission made clear in Order No. 670 that a duty to speak to avoid making untrue statements or material omissions would arise only as a result of a tariff or Commission order, rule or regulation. However, the Anti-Manipulation Rule extends to situations where an entity has either voluntarily or pursuant to a tariff or Commission directive provided information but then misrepresents or omits a material fact such that the information provided is materially misleading.

A fact is material if there is a substantial likelihood that a reasonable market participant would consider it in making its decision to buy or sell because the material fact significantly altered the total mix of information available. To violate the Anti-Manipulation Rule, one must act with a suffi cient state of mind, that is, level of intent. As with violations of SEC Rule 10b-5, intentional conduct or recklessness (known as scienter) are enough to satisfy the Rule. The Commission also stated that for conduct to violate the "in connection with" element of the Rule, there must be a suffi cient nexus between an entity's fraudulent conduct and a jurisdictional transaction. In committing fraud, the entity must have intended to affect, or have acted recklessly to affect, a jurisdictional transaction.

In: Natural Gas and Electricity
Editor: Lee Huang

ISBN: 978-1-62417-712-5

Chapter 2

NATURAL GAS IN THE U.S. ECONOMY: OPPORTUNITIES FOR GROWTH*

Robert Pirog and Michael Ratner

SUMMARY

Due to the growth in natural gas production, primarily from shale gas, the United States is benefitting from some of the lowest prices for natural gas in the world and faces the question of how to best use this resource.

Different segments of the U.S. economy have different perspectives on the role natural gas can play. Suppliers, which have become the victims of their own production success, are facing low prices that are forecast to remain low. Some companies that have traditionally produced only natural gas have even turned their attention to oil in order to improve their financial situation. Smaller companies are having a difficult time continuing operations and larger companies, including international companies, have bought into many shale gas assets. Prices have remained low even as consumption has increased, in part, because producers have raised production to meet the demand and because companies have improved efficiency and extraction techniques. Some companies, many with large production operations, have applied for permits to export natural gas. This has raised concerns from consumers of natural gas that domestic prices will rise. The debate regarding exports is ongoing.

Industries that consume natural gas have seen input costs drop, and some have heralded low natural gas prices as the impetus for a manufacturing revolution in the United States. Some companies have begun to make major investments to take advantage of the low natural gas prices, particularly in petrochemicals. Other companies are waiting to see if prices will remain low long enough to warrant major investments in new facilities. Meanwhile, the electric power sector has already seen a transition from coal-fired generation to natural gas. Low natural gas prices are also putting pressure on renewable sources of power generation. However, increases in demand will put upward pressure on natural gas prices.

The transportation sector, the one part of the economy vulnerable to foreign energy supplies, is beginning to explore ways to use more natural gas. Transportation makes up

* This is an edited, reformatted and augmented version of Congressional Research Service, Publication No. R42814, dated November 6, 2012.

less than 1% of U.S. natural gas consumption and would require billions of dollars in investment to increase that share significantly.

All of the change that has taken place so far has occurred despite environmental concerns and regulatory developments at the state and federal level that might curtail production. Natural gas is a fossil fuel that produces various pollutants, some more than other fossil fuels and some less. Methane, the major component of natural gas, is also a potent greenhouse gas when released without burning. Other environmental concerns focus on water use and disposal in hydraulic fracturing to extract natural gas from shale formations.

Over the next five years, many of the issues being debated now may be decided. The industry and market are adapting to the newly found supplies and the concerns associated with them, as well as integrating more natural gas into the economy. There are many evolving issues some of which Congress can influence directly because of statutes and some indirectly. On the demand side, legislation has been introduced regarding exports of liquefied natural gas and alternative fuels for vehicles. There has been other legislation related to environmental regulations of natural gas.

Introduction: What To Do With All the Natural Gas?

The relatively rapid expansion of U.S. natural gas resources over the last five years, particularly from shale gas, has been coupled with slower demand growth by natural gas consumers. The result has been low prices not seen for over a decade and, equally important, prices that are projected to stay low for decades. U.S. natural gas prices have also been comparatively lower than those observed in international markets. The projected persistence of rising supply and low prices has raised the question of how the United States will take advantage of its natural gas resources. Because of low prices, there have been requests by some producers to export natural gas in liquefied form (LNG), hoping to sell at higher world prices. Some consumer groups argue that exports will raise domestic prices, which will hurt businesses and households. Some businesses believe the low prices can spur a resurgence in U.S. manufacturing, particularly petrochemicals and other industries using large amounts of natural gas. Some environmentalists view natural gas as a key component for decreasing carbon dioxide (CO_2) emissions and other greenhouse gases, while other environmentalists believe that it is as polluting as coal, especially when derived from shale formations. Some analysts believe natural gas can enhance U.S. energy security if it can be used in transportation to replace gasoline. All these questions highlight the tremendous changes that have taken place in the U.S. energy landscape over the last few years and portend future changes. There have been over 150 bills introduced in the 112th Congress that would affect both supply of natural gas (H.R. 840, S. 302, S. 706, and S. 1007, among others) and demand, with some targeting specific sectors such as LNG exports (H.R. 3913 and H.R. 4024, among others) and transportation (S. 734 and H.R. 970, among others). Legislation has also been introduced that would affect environmental issues associated with natural gas (H.R. 1084, H.R. 4322, and H.R. 6235, among others). This report examines what has changed in the natural gas industry and focuses on the demand side and ancillary benefits to the U.S. economy. This report does not address negative consequences to other economic sectors due to a shift to natural gas.

BACKGROUND: THE MARKET HAS CHANGED

Shale Gas: The Game Changer

Without the development of shale gas, much of the discussion about integrating more natural gas into the economy would not be occurring. Over the last decade U.S. natural gas reserves have climbed tremendously, 72% since 2000 and 49% since 2005 (see Figure 1).[1] These data allow for reductions for natural gas produced during the period and so are "net" increases. In recent years, the increase in reserves is mostly attributed to development of shale gas, which has grown from 10% of U.S. natural gas reserves in 2007 to 32% in 2010. By comparison conventional U.S. natural gas reserves declined between 2007 and 2008, and fell again in 2010. Though the decline was marginal, it highlights the importance of shale gas to future U.S. natural gas production. Many industry analysts expect shale gas reserves to continue to rise and make up a greater portion of U.S. natural gas reserves unless new restrictions are placed on the industry, such as related to hydraulic fracturing, power plant emissions, etc.

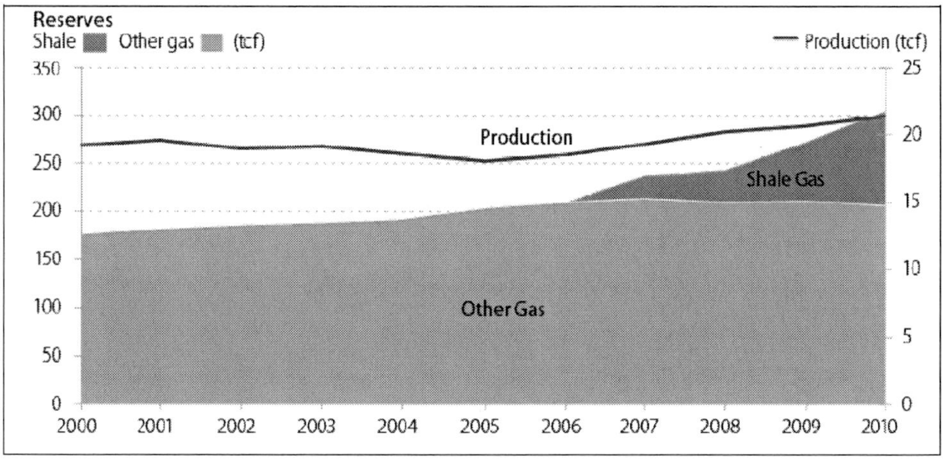

Source: U.S. Energy Information Administration—http://www.eia.gov/dnav/ ng/ ng_enr_ dry_a_EPG0_ r11_bcf_a.htm, http://www.eia.gov/dnav/ng/ng_enr_ shalegas_ a_EPG0_R5301_ Bcf_a.htm, and http://www.eia.gov/dnav/ng/hist/ n9070 us2a.htm.

Notes: Although shale gas reserves existed prior to 2007, they were not as significant a portion of overall reserves and not tabulated prior to 2007. 2010 is the latest year data are available.

Figure 1. U.S. Natural Gas Reserves and Production, 2000 – 2010.

In 2011, the United States produced and consumed more natural gas than it ever has—23 trillion cubic feet (tcf) and 24 tcf, respectively—while paying some of the lowest market prices for natural gas in the world.[2] The production figure of 23 tcf is of dry gas, which has been processed for consumption purposes, but the United States actually produced 28.5 tcf of raw natural gas in 2011. The United States is the world's leading producer of natural gas, surpassing Russia in 2009, and the world's leading consumer. After declining for the first half of the last decade, U.S. natural gas production rose 18% in the latter half, with shale gas accounting for 25% of production by 2010.[3]

Reserves and production data do not tell the whole story when looking at the U.S. transformation regarding natural gas supply. The term *reserves* has a specific industry definition that includes a technological component, an economic factor, and a probability of success among other criteria. To more fully understand the changes to the U.S. natural gas sector it is more appropriate to look at reserves and estimates for undiscovered, technically recoverable resources (UTRR) (see **Figure 2**). UTRR is an estimate of what can be extracted using current technology regardless of price. Using UTRR plus reserves, the United States has a natural gas resource base of 1,809 tcf or enough gas for approximately 79 years of production at 2011 levels. Compared with data from 2006, U.S. UTRR for natural gas has jumped almost 25%. Even this measure may not accurately reflect what will be extracted from the ground as technology is constantly changing. Just over the last few years, industry has been able to improve its shale gas extraction rate from about 5% to about 15%, thereby tripling what is recoverable.

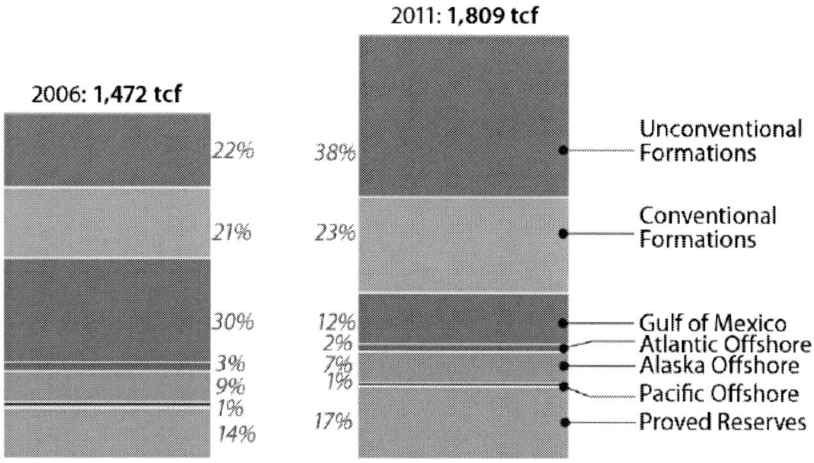

Source: Department of the Interior's U.S. Geological Survey and Bureau of Ocean Energy Management (BOEM) and U.S. Energy Information Administration.

Notes: The data for unconventional includes some but not all of the shale basins as some have not been assessed to date by the U.S. Geological Survey. Undiscovered technically recoverable resources (UTRR) refers to amounts of natural gas estimated to exist by examining geologic characteristics of unexplored areas and recoverable using current technology. All the figures in the graphic above are UTRR except Proved Reserves, which is defined as a 90% probability of recovery using existing technology and at current prices. Units = trillion cubic feet (tcf).

Figure 2. Natural Gas Resources and Reserves, 2006 vs. 2011.

Projected Future Growth

In 2011, natural gas was the most produced fuel, on a tonne of oil equivalent basis, in the United States, surpassing coal for the first time. This change was driven by the success of shale gas development. The U.S. Energy Information Administration (EIA), which makes projections based on current policy and information, estimated in its *2012 Annual Energy Outlook* that overall U.S. natural gas production will grow 28% between 2010 and 2035 (see **Table 1**).

Table 1. U.S. Natural Gas Production Composition, Imports & Prices Projections through 2035, tcf unless noted

	2010	2010 % of Total	2015	2020	2025	2030	2035	2035 % of Total
Associated Gas (Onshore)	1.4	6.5%	1.5	1.5	1.4	1.2	1.0	3.6%
Tight Gas	5.7	26.3%	6.1	6.1	6.2	6.1	6.1	21.9%
Shale Gas	**5.0**	**23.0%**	**8.2**	**9.7**	**11.3**	**12.4**	**13.6**	**48.9%**
Coalbed Methane	2.0	9.2%	1.8	1.8	1.8	1.7	1.8	6.5%
Other Non-Associated Gas	4.6	21.2%	3.8	3.4	3.0	2.7	2.4	8.6%
Lower 48 Offshore	2.6	12.0%	1.9	2.3	2.4	2.6	2.7	9.7%
Alaska	0.4	1.8%	0.3	0.3	0.3	0.3	0.2	0.7%
Total Production	21.7	100%	23.6	25.1	26.4	27.0	27.8	99.9%
Net Imports	2.6		1.7	0.4	(0.8)	(0.9)	(1.4)	
Spot Prices ($/MBtu)	$4.39		$4.29	$4.58	$5.63	$6.29	$7.37	

Source: EIA Annual Energy Outlook 2012, Reference Case, http://www.eia.gov/oiaf/aeo/tablebrowser/#release=AEO2012&subject=0-AEO2012&table=14-AEO2012&region=0-0&cases=ref2012-d020112c.

Notes: Data for 2010 are actual figures, but data going forward are projections by EIA under their Reference Case for the 2012 Annual Energy Outlook. Volume units = trillion cubic feet.

Shale gas will comprise almost 49% of that production, up from 23% in 2010. During the time period the United States is expected to go from a net importer of natural gas by pipeline and LNG to a net exporter by 2022, which is a change from EIA's *2011 Annual Energy Outlook* when there was no time period in which the United States was forecast to be a net exporter of natural gas. The United States is forecast to be a net LNG exporter by 2016, according to EIA.

The only portion of U.S. production to significantly rise over the time period examined by EIA is shale gas. It grows both as a percent of overall production as well as in absolute terms. As a whole, conventional natural gas is estimated to make up a much smaller percent of the overall production mix, declining from 41.5% in 2010 to 22.7% in 2035. The decline of conventional natural gas sources demonstrates that natural gas is a finite resource, but how much may exist is still unknown.

Natural Gas Prices: A Competitive Advantage

A consequence of the rapid increase in natural gas supply is downward pressure on prices. U.S. spot natural gas prices, also known as the Henry Hub price or the NYMEX (New

York Mercantile Exchange) price, are relatively low compared with domestic prices over the last decade as well as international prices over the last few years (see Figure 3 and Figure 5).

Historically, natural gas prices in the United States have been volatile. From 1995 to 1999 the spot price of natural gas averaged $2.23 per MBtu, but increased to an average price of $4.68/MBtu during the 2000 to 2004 period, an almost 110% rise. From 2005 to 2009 the spot price averaged $7.23/MBtu hitting a peak of $15.38/MBtu in December 2005. Prices again spiked in July 2008. High prices, based on an assumption of a declining reserve base and production, seemed to indicate that the future would be characterized by much higher natural gas prices and domestic exposure to high-priced, imported LNG. However, prices started to decline in the latter half of 2008 because of the recession. As demand picked up in the 2010 to 2011 period, along with a warm winter, the average spot price of natural gas did not, declining to $4.20/MBtu. In the first eight months of 2012 spot prices averaged $2.55/MBtu.[4] These lower recent prices and optimistic expectations concerning domestic supply have led to a view that the United States will have plentiful supplies of natural gas available at low cost well into the future. Nevertheless, some industry participants have experienced price trends that reversed in the past and not all are ready to make financial commitments despite current low prices. As mentioned previously, shale gas production began coming to the market in 2007 and has been increasing ever since.

Source: Bloomberg New Energy Finance.

Notes: Units = dollars per million British thermal unit ($/MBtu).

Figure 3. U.S. Natural Gas Prices, 1990-2012.

The price of natural gas, as with other commodities, is driven by supply and demand. Demand for natural gas is affected by the economy, the weather, seasonality, product storage, and electrical use, among other factors (see **Figure 4**). The winter months in the United States are when demand for natural gas peaks, as it is used in residential and commercial buildings for heating. The United States also has a second, more modest peak, in the summer as more natural gas is consumed for electricity generation for air conditioning. This aspect of the demand cycle will vary depending upon the region of the country. During the months of approximately April through June, natural gas is normally produced and stored for the winter peak demand starting in November. As a consequence, storage plays a critical role in balancing the U.S. natural gas market and can greatly affect prices.

As an example, in Spring 2011 large underground storage facilities were being filled as typically happens, but the winter months later in the year were warmer than expected and storage levels remained high as less natural gas was used for heating. In the Spring of 2012, the high storage levels added additional downward pressure on prices, which dropped to about $2.00/MBtu for the first time since the 1990s. Producers of natural gas tend not to stop production from a well, unless prices are below their operating costs and they believe they will stay at that level for a long time, and even then companies may be hesitant to stop production. There are costs and technical issues involved with halting production, maintaining a well, and restarting production.

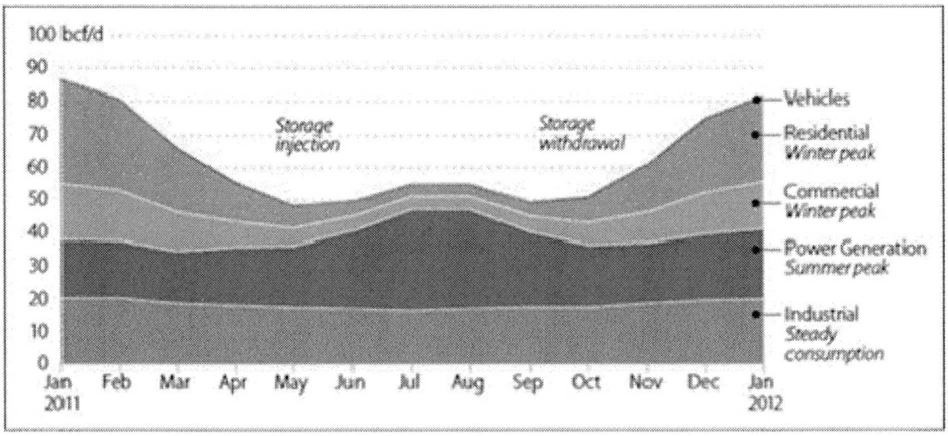

Source: U.S. Energy Information Administration (EIA) and Bloomberg New Energy Finance.
Notes: Units = billion cubic feet per day (bcf/d).

Figure 4. Seasonal Natural Gas Demand.

The advent of shale gas and the decline of U.S. natural gas prices has attracted global attention and prompted countries to try to emulate the U.S. success in developing their unconventional gas resources. Although other countries have touted their unconventional natural gas resources, no country has achieved the level of development of the United States, except Australia, in developing their coal seam gas. Canada is moving ahead with its shale gas development, but lags behind the United States. As can be seen in **Figure 5**, U.S. and other regional natural gas prices around the world moved in sync for most of the last decade even though there is not a global market for natural gas as there is with oil. From 2008 to 2009, natural gas prices dropped worldwide because of the decrease in demand from the decline in economic activity. U.S. shale gas was beginning to come to market in 2007/2008 and by 2010/2011 it changed the trajectory of U.S. natural gas prices from those of the rest of the world. In 2011, the rest of world faced higher prices than in 2010 for natural gas, but the United States saw its natural gas price decline by 9%. U.S. natural gas prices have continued to trend lower ever since, and many analysts forecast U.S. natural gas prices to remain relatively low at least through the end of this decade and possibly for longer.

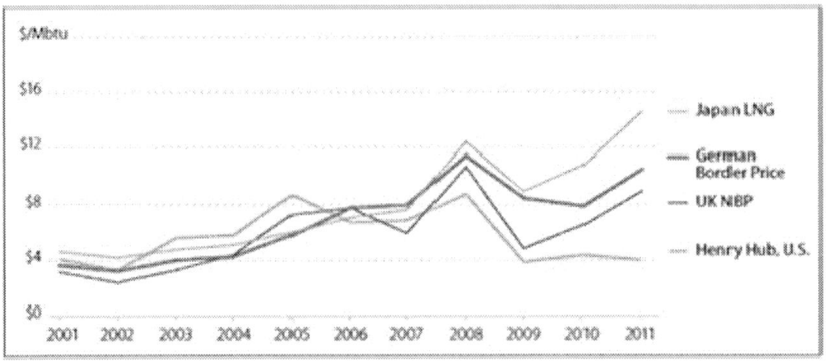

Source: BP Statistical Review of World Energy, 2012, June 2012, p. 27. http://www.bp.com/sectionbodycopy.do? categoryId=7500&contentId=7068481.

Notes: The German Border Price is a proxy for European oil indexed prices. Units = U.S. dollar per million British thermal units (MBtu).

Figure 5. Select Regional Natural Gas Prices, 2001-2011.

The differential between the U.S. spot price and the average for the other major natural gas markets has been growing each year since 2007. Although there are many factors that can affect natural gas prices in a particular market for a certain period, such as the nuclear accident in Japan, the growing differential highlights how the U.S. natural gas market has been insulated from external events and the impact of the expanding U.S. natural gas resource base.

Contributing Factors

As natural gas prices in the United States have declined other market factors have begun to alter the industry and affect market fundamentals. In order to improve their revenues, many natural gas-oriented companies have added more oil-rich areas to their portfolios, particularly tight-oil formations, which have contributed to growing U.S. oil production.[5] However, with oil becoming the focus of production, natural gas infrastructure is not keeping pace and consequently natural gas is being flared in some areas of the United States in large quantities. Natural gas use displacing coal in electric generation has already shown the potential for decreasing emissions, as natural gas power generation releases about half the CO2 of coal-fired generation. Additionally, companies are seeking further markets for natural gas and have applied for permits to export natural gas as LNG. As highlighted in Figure 5, the price for natural gas in the United States is lower than other major markets.

Natural Gas Liquids: A Production Driver

Natural gas liquids (NGLs) have taken on a new prominence as shale gas production has increased and prices have fallen. NGL is a general term for all liquid products separated from natural gas at a gas processing plant and includes ethane, propane, butane, and pentanes. When NGLs are present with methane, which is the primary component of natural gas, the natural gas is referred to as either "hot" or "wet" gas. Once the NGLs are removed from the methane the natural gas is referred to as "dry" gas, which is what most consumers use. Each

NGL has its own market and its own value. As the price for dry gas has dropped because of the increase in supply and other reasons such as the warm winter of 2011, the natural gas industry has turned its attention to producing more wet gas in order to bolster the value they receive. Some companies have shifted their production portfolios to tight oil formations, such as the Bakken in North Dakota, to capitalize on the experience they gained in shale gas development.

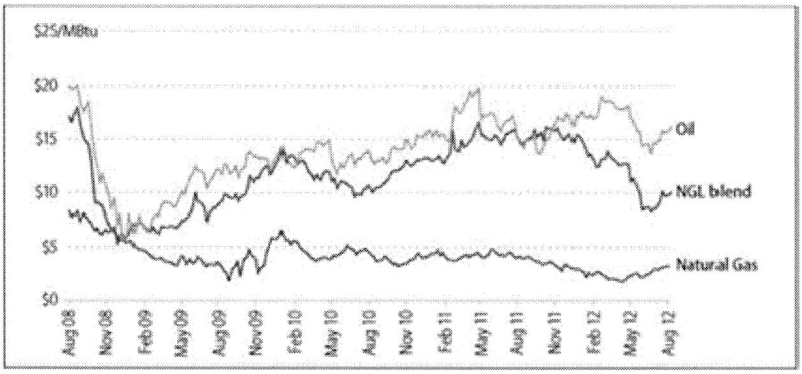

Source: Bloomberg New Energy Finance.

Notes: The NGL blend prices are weighted to an average mix of 42% ethane, 29% propane, 11% normal butane, 6% iso-butane, and 13% natural gasoline and priced at Mont Belvieu, TX. The mix of NGLs will vary by source and the price will vary by the actual market for the product. The natural gas price is at Henry Hub and the oil price is West Texas Intermediate (WTI). August 2008 is when prices for all three commodities just started to decline from their peaks in July because of the recession. Units = million British thermal units (MBtu).

Figure 6. Natural Gas, Oil, and NGL Prices, August 2008 to August 2012.

Historically, the individual NGL products have been priced against oil, and as oil prices have remained higher since 2005 relative to natural gas, it has driven an increase of wet gas production, thereby maintaining the amount of dry gas as a production "byproduct" despite its low price.

As can be seen in **Figure 6**, the price of natural gas at the end of August 2012 was approximately $3.20/MBtu, while the value of NGLs was almost $10.00/MBtu. The additional value that can be extracted above the price of natural gas is a driver of current production. Drilling rigs are being moved from dry gas fields to wet gas and predominantly oil fields, like the Bakken, that produce natural gas in association with oil production. The proportion and composition of NGLs varies by gas field and therefore the price differential will vary as well. The price differential between oil and NGLs is also diverging over the last year because the prices of ethane and propane, two main components of NGLs, are delinking from their historic connection to oil prices. As ethane and propane production has risen the last couple of years their prices have been driven more by their own market fundamentals of supply and demand.

Flaring: A Value Issue

As natural gas production began to increase in 2005, so did the amount of natural gas that was vented (released into the atmosphere) and flared (burned at the production site). Natural gas is either flared, which is preferred, or vented usually for safety or health reasons prior to connecting a well to a pipeline. This issue has grown in prominence as development of tight oil resources has increased because of the horizontal drilling and hydraulic fracturing techniques developed for shale gas has shifted to that sector. The Bakken formation has had the most notoriety because as its oil production has increased, so has the amount of natural gas that is being flared (over 20% of gross production in 2010).[6] Oil production in tight formations is expanding rapidly and the natural gas infrastructure to move the associated gas to market has lagged. The low price of natural gas has compounded the issue, particularly in jurisdictions that have relatively lax regulations when it comes to flaring.

Factors Affecting Production

The location of shale formations (see **Figure A-1**) has altered the movement of natural gas supplies to consuming markets in the United States. As an example, traditionally gas flowed from the Gulf Coast to the Northeast via large diameter, long-haul transmission pipelines. The discovery and production from the Marcellus Shale formation, which underlies much of West Virginia and Pennsylvania, southern New York, eastern Ohio, western Maryland, and western Virginia, has changed the need for gas from the Gulf Coast as the Marcellus is much closer to Northeast markets.[7] The reconfiguration of supply centers and consuming markets requires new infrastructure to be put in place, including gathering pipelines (small diameter pipelines that bring gas from the field to processing facilities), transmission pipelines, and processing facilities (plants that remove other hydrocarbons, i.e., NGLs, carbon dioxide, water, and nitrogen among other things before transmission and consumption). Additionally, other transmission pipelines will be reversed in order to bring the natural gas to different locations. This reconfiguration is an ongoing process that may cost the industry billions of dollars to achieve greater efficiencies.

Water is a key component of energy production, particularly for shale gas development that requires significant quantities for hydraulic fracturing.[8] Unfortunately, there are significant data gaps in this area that make evaluating the water needs of the energy sector difficult. A fracture treatment for a single zone may use more than 500,000 gallons of water.[9] Wells that need multiple treatments usually require between 3 million to 5 million gallons or more. For comparison, an Olympic-size swimming pool holds over 660,000 gallons of water and the average daily per capita consumption of fresh water (roughly 1,430 gallons per day) works out to 522,000 gallons over one year.[10]

Historical Natural Gas Use

The transition to using more natural gas in the economy is already underway. Natural gas comprised 28% of the U.S. primary energy mix in 2011 and is on the upswing, while oil and coal have both declined in absolute terms and as a percent of consumption over the last decade. Historically, natural gas is well below its primary energy market share high of 34% in 1971, but just above its 20-year average of 26% (see **Figure 7**).

Natural gas consumption, which is in part tied to economic activity, is on the rise in the United States after declining in 2009 because of the recession. Unlike oil, which is mostly consumed in transportation, and coal, which is mostly consumed in electricity generation, natural gas has more diversity to its consumption. Almost a third each of natural gas consumption is from electricity generation, industrial, and residential/commercial, which tends to use it similarly (see **Figure 8**). EIA divides consumption into five economic categories—residential, commercial, industrial, electric power, and transportation (vehicle fuel)—and two operational categories, lease and plant fuel (gas used for production and at processing plants) and pipeline and distribution use (gas used for compressors and to move gas to consumers). The two operational categories are usually not included when discussing U.S. gas consumption.

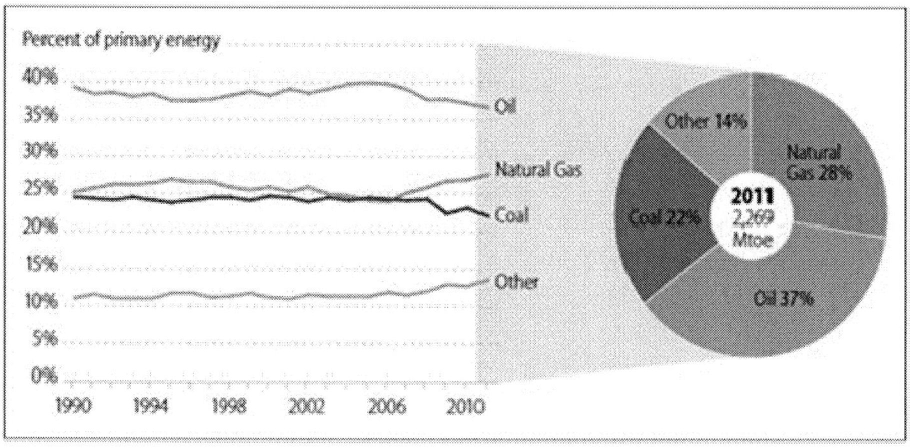

Source: BP Statistical Review of World Energy, 2012, June 2012. http://www.bp.com/sectiongenericarticle800.do?categoryId=9037130&contentId=7068669.

Notes: Other primary energy includes nuclear, hydroelectricity and renewables. Units = million tonnes of oil equivalent (Mtoe).

Figure 7. U.S. Primary Energy Consumption, 1990-2011.

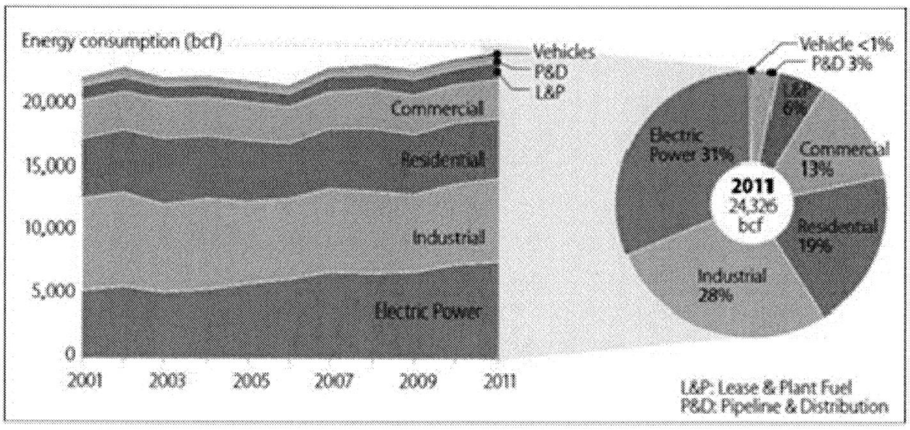

Source: U.S. Energy Information Administration, http://www.eia.gov/dnav/ng/ng_ cons_sum_ dcu_ nus_a.htm.

Notes: Units = billion cubic feet (bcf).

Figure 8. U.S. Natural Gas Consumption by Sector, 2001-2011.

All sectors increased in 2011 over 2010 figures, except residential consumption, which decreased, mainly due to warmer weather. There is limited additional residential and commercial use of natural gas unless new regional markets can be opened, which will require investment and infrastructure.

lectric power generation makes up the biggest component of U.S. natural gas consumption followed by industrial use. (See sections "Electric Power Generation: First Mover" and "Petrochemicals: A Possible Rejuvenation.") However, the two sectors have essentially reversed position over the last decade, with electric power growing by about 10% since 2001 and industrial consumption declining by almost the same amount.

The reversal of the electric power and industrial sectors can be attributed to rising prices early in the decade that decreased industrial use and prompted much of the sector to move overseas.

Electric generation from natural gas also decreased during the early part of the decade, but when prices started to decline because of the lower demand and then increased supply, natural gas-fired generation was the beneficiary. Some of the industrial sector is poised to return to the United States from overseas because of projected low natural gas prices, although in many cases companies are waiting to determine if low U.S. natural gas prices are sustainable.

In April 2012, for the first time in history, the amount of electricity generation from natural gas equaled that of coal, according to EIA statistics, each with about 32% of the market.[11]

As mentioned above, natural gas use in electric generation has been growing over the previous decade, and will likely continue to grow, particularly if natural gas prices remain low. In addition, concerns about more stringent environmental regulations have in some cases contributed to the increase in retirement of coal-fired electricity generation, which largely is base-load capacity and is being replaced by natural gas generation.

An apparent consequence of the shift to more natural gas-fired electric power generation in the first quarter of 2012 has been a decrease of U.S. carbon dioxide emissions by almost 8% to their lowest levels since 1992.[12] Besides more natural gas-fired electric power, EIA attributes the decline in emissions to reduced household heating because of the mild winter and reduced gasoline demand.

Figure 9 also highlights the relative price variability of each commodity, with natural gas showing a much wider range than coal during the time period, as well as a significant decrease in the spread between natural gas and coal prices.

If large parts of the U.S. economy are to shift to natural gas, sufficiently low long-term prices that maintain the advantage of gas over other fuels are likely to be required. However, the increases in demand associated with these sectoral shifts in favor of gas use could result in prices also increasing, perhaps bringing into question the economic advantages available to potential users of natural gas.

Further, if prices were to remain low in the long-term, industry might not be able to sustain production. If wellhead prices are too low for developers to make sufficient profits, exploration and development activities might slow and production might be capped. Market forces will likely establish a workable balance between the demand and supply sides of the market.

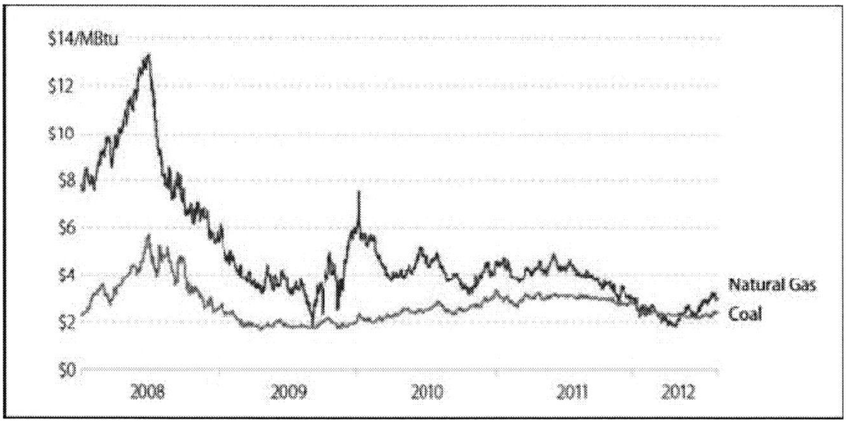

Source: Bloomberg New Energy Finance.
Notes: Units = dollars per million British thermal unit ($/MBtu).

Figure 9. U.S. Natural Gas and Coal Prices, 2008-2012.

NATURAL GAS MARKETS: THE POSSIBILITIES

If natural gas were able to replace oil and coal in the fuel mix, it would require almost 77 tcf of natural gas per year; approximately tripling current U.S. consumption and production (see **Figure 10**). At that rate, even the current estimated resource base of 1,809 tcf would be exhausted in about 23 years. The amount of investment that would be required to undertake this change would be enormous, probably in the trillions of dollars, and new technologies and infrastructure would be needed to make the changes economical and practical, particularly in the transportation sector for natural gas vehicles.

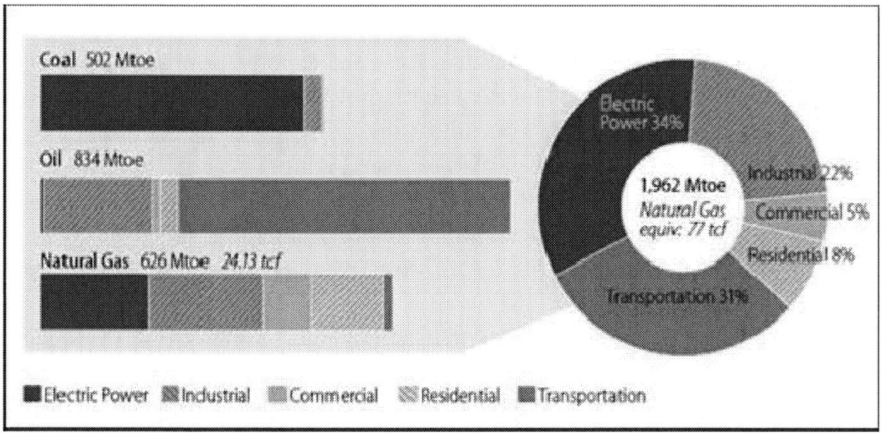

Data Source: BP Statistical Review of World Energy, 2012, and the U.S. Energy Information Administration.
Notes: Units = million tonnes of oil equivalent (Mtoe), which equates to 39.2 billion cubic feet.

Figure 10. 2011 Hydrocarbon Disposition, Natural gas requirements by sector.

Almost 24 tcf of natural gas would be required in the transportation sector alone, which is approximately the amount the United States consumed in total in 2011. Less than 1% of U.S. natural gas was consumed in the transportation sector in 2011. Although this scenario is beyond what could be achieved realistically, it highlights the central questions of this report as to how much natural gas the United States can produce and what sectors of the economy can use it.

Demand Response: Direct Beneficiaries

Expanded supply, coupled with low natural gas prices, has the potential to contribute to a transformation of important sectors of the U.S. economy. Increased output and employment, expanded investment, income growth, improved competitiveness, and a reduction in the foreign trade deficit are likely outcomes. These conditions in the natural gas markets are likely to benefit certain key industries directly, while many other industries could experience indirect benefits. Direct beneficiaries are those industries that use natural gas as a raw material or as an important input in a production process. Industries whose output is directly related to the expansion of natural gas exploration, development and production are also direct beneficiaries. Examples of industries that use natural gas directly are petrochemicals and fertilizers. The steel industry is an example of an industry whose output is linked to the pace of natural gas resource development. Industries experiencing indirect benefits might include construction and capital goods producers that contribute to the supply chain for the investment projects undertaken by expanding natural gas consumers. In addition, more spending by workers in all of these industries could increase the growth of a wide variety of consumer goods and retail firms. The economic benefits of shale gas development and production will also open areas not recently accustomed to natural gas production, for example, the Marcellus field in parts of Pennsylvania, Ohio, West Virginia, Maryland, Virginia, and New York.

In the international economy, those U.S. industries directly affected by expanded supply and low natural gas prices are likely to experience a competitive advantage over the producers of similar goods in other countries, resulting in increased exports from, and decreased imports to the United States. These effects would likely improve the U.S. trade deficit position. This advantage is likely to be maintained over time if the U.S. price of natural gas remains below those observed in other world regional markets (see Figure 5).[13] U.S. industry's advantage could be reduced through a process of world natural gas price convergence, especially in the three leading regional markets. However, for this to occur, traditional long-run contract terms, specifically linking natural gas prices to oil prices, would need to be changed to a more market-oriented method.

While low natural gas prices are likely to benefit many consuming industries, they have a depressing effect on the natural gas industry itself. If prices fall, and remain below critical levels, the industry could respond by reducing exploration and development activity, and/or capping existing wells. Although some of this has occurred, overall natural gas production continues to trend upwards, in part because of cost-reducing efficiency gains, and in part because of contract provisions.[14] Actions that reduce current and future supplies tend to cause the price of gas to rise. One mitigating factor to this price process is the extraction of natural gas liquids (NGLs) from wells (see "Natural Gas Liquids: A Production Driver"). These liquids are typically of high value and may validate the production of natural gas even in a low price environment for the natural gas itself, which assumes oil prices remain high. Also, for some industries, the ratio of the price of oil to the price of natural gas is more important to U.S. competitiveness than the absolute price of domestic natural gas. An example is petrochemicals, where U.S. producers use natural gas liquids as a feedstock for ethylene production while European producers tend to use naphtha, derived from crude oil. Increased production of ethane from shale gas development has given the U.S. petrochemicals industry a price advantage.

Electric Power Generation: First Mover[15]

As discussed earlier in this report, natural gas is expected to increase its share of electricity generating capacity. The broad reasons for this increase begin with the expected increases in natural gas supply, along with low prices. Also important are the relatively lower carbon emissions of gas-fired plants relative to coal-fired power plants, and the relatively high capital investment costs of coal-fired plants compared to natural gas-fired plants.[16] An additional benefit for the power-generating industry related to natural gas-fired plants is flexibility. Natural gas facilities can increase or decrease generation much more efficiently and cheaply than coal-fired plants. Much of the increase in natural gas-fired generation has been from facilities that have been operating below capacity. Lower fuel costs have given these facilities an advantage over other generators.

As a result of these advantages natural gas-fired plants are expected to account for about 60% of new generating capacity in the United States between 2011 and 2035.[17] Job creation as a result of this shift to natural gas is likely to be small. Although construction and capital equipment jobs will be created as natural gas-fired power plants are constructed, these jobs are not permanent. The net number of jobs in electric power generating facilities is likely to be low because jobs in coal fired-plants may be lost as new gas-fueled plants begin operating. In addition, a reduced demand for coal from the electric power generating sector may result in mining jobs shrinking, while gas drilling jobs go up.

Coal will, however, retain an important share of the electric power market. Some large facilities have scrubbing equipment in place to reduce emissions and many large producers are located near coal mines, which insures low transportation costs. In contrast, in many cases new natural gas pipelines would have to be constructed to allow switching from coal to natural gas.[18] Additionally, should natural gas prices rise because of increased demand in electricity or other sectors, coal-fired generation will become more competitive again.

The expansion of natural gas as a fuel for electricity generation is expected to have important macroeconomic effects, even though direct job creation is not likely to be great. The reason for this is based on the cost structure of electric power generation, where fuel costs account for approximately 40% of total production costs. If the cost savings relative to natural gas use materialize, and if they are passed on to consumers in the form of lower electricity rates, household disposable income would increase. This increase could be used by households to finance the purchase of a wide variety of consumer goods. Commercial natural gas consumers could reap higher profits, or contain growth in the cost of goods sold, due to lower electricity costs. Industrial demand could experience similar benefits; however, the effects are likely to be less important as the percentage of total costs accounted for by electricity costs falls. The exception is the petrochemicals industry, where natural gas costs are a large part of total costs.

Petrochemicals: A Possible Rejuvenation

Natural gas is used in the chemicals industry both as a fuel and as a raw material. As stated previously, when natural gas is produced it is mainly methane, but also may include NGLs. One NGL is ethane, which is used to produce ethylene. Ethylene, a key component in plastics, in turn, is used to produce a wide variety of consumer goods, ranging from food packaging to home siding and window frames, automotive anti-freeze, clothing, tires, and bottles.

In 2012, a number of chemical companies announced plans to invest in new plant capacity, expand existing facilities, or re-open plants near shale gas supplies. In April 2012, Dow Chemical announced $4 billion in expansions and new investment in Texas. Shell Chemical announced plans for an ethane cracking unit costing between $2 billion and $4 billion, to be constructed in Pennsylvania near Marcellus Shale natural gas supplies. Chevron announced plans for a $1 billion investment at its Baytown facility in Texas. In addition, Phillips Chemical, Westlake Chemical, and others announced investment plans related to low-cost shale gas availability.[19]

Many of these investment plans, if they ultimately come to fruition, will result in new production capacity becoming available over the next five years. Multi-billion dollar investments with time lags of up to five years before they generate cash flow suggest that the companies believe that the United States is entering a period of sustained low natural gas prices and growing supply. Without supply growth, the increase in demand from these facilities would likely push prices up. The history of natural gas prices in the United States, as shown in **Figure 3** demonstrates that neither high, nor low, gas price stability has been evident in the market for long periods of time.

The American Chemistry Council conducted a study of the effects on the U.S. economy of a onetime $16.2 billion investment program, over several years, in petrochemical plant and equipment related to low-cost natural gas availability.[20] Their assumption was that this level of investment would increase ethane capacity in the United States by 25%.[21] Their analyses determined that 17,000 new, relatively high paying jobs would be created in the chemical industry, along with 395,000 jobs outside the chemical industry. Tax revenues generated as a result of the investment program would be about $4.4 billion annually for federal, state and local governments. Output of the U.S. chemicals industry could increase by $32.8 billion and a total of $132.4 billion in output would be generated by the economy as a whole. Studies of this type assume various economic and other conditions which, if they should change, would also change the result. While the specific forecast values of studies estimating economic effects are unlikely to be precisely correct over time, they may indicate potential and order of magnitude.

It has been reported that the global ethylene production balance between oil-based naphtha and natural gas-based ethane has been about 80:20 in favor of naphtha. However, the balance has tipped to 50:50, even before the large U.S. investments in ethane capacity have come on stream. This shift in input utilization has had negative effects on the international naphtha market, both for the U.S. and foreign markets. U.S. refiners report an excess of naphtha, much of which is appearing in the European market, causing a supply glut and falling prices. European refiners, experiencing poor economics due to the euro crisis and the onset of recession, are suffering further losses due to the new competition from the United States. The normally high price of naphtha, comparable to that of gasoline, makes the competition more economically damaging for European refiners.[22]

The Fertilizer Industry Could Help Farmers

Natural gas is the primary raw material in nitrogen-based fertilizer production. From 70% to 90% of the estimated cost of producing nitrogen-based fertilizers is related to the cost of natural gas. In the 2000s, when natural gas prices for industrial consumers more than doubled, closure of ammonia plants, which supply the raw material for fertilizers, followed suit, rising from 13 in 2002 to 26 in 2007. While some of this capacity moved overseas, some was permanently closed. It would take time and investment to reinstate U.S. capacity. If U.S. fertilizer production could recover and pass on lower costs to farmers, this could lower the cost of food and ethanol for use in transportation, and have employment benefits in the fertilizer industry as well as those industries whose costs had decreased.

In the short-run few of these economic benefits can be observed. The fertilizer industry appears to be taking a wait-and-see attitude with respect to natural gas prices before major investment decisions in capacity expansion are undertaken. The industry, which suffered from high and volatile natural gas prices in the U.S. market, is likely to need compelling evidence that gas supplies will remain abundant and prices low in the future. For now, U.S. fertilizer cost savings due to low natural gas prices are balanced in favor of increased profits for producers compared to cost savings for consumers. This condition is encouraged by the rapidly increasing demand for corn which is keeping the demand for fertilizers high. For the period 2000 to 2006, the average acreage of planted corn in the United States was 79 million. In 2011 the total increased to 92 million, an increase of 16%, and the total is expected to continue rising.[23]

Steel Production: Two-Fold Winner

The expansion of shale gas and tight oil exploration and development has had an expansionary effect on the U.S. steel industry. Product demand has increased, while operating costs have declined. Natural gas exploration generates demand for tubular goods used for pipes, tubes, and joints in gas drilling equipment. In addition, natural gas is being substituted for coal as an input in the steel industry's blast furnaces, driving down costs per ton.[24]

Steel facilities located in, or near, areas of shale gas development are experiencing growth. For example, the steel industry in Ohio has been reported to be in the process of adding some two million feet of production space, costing $1.5 billion.[25] A number of companies, including Timken, United States Steel, Vallourec & Mannesmann, and a United States Steel-Kobe Steel joint venture project are located in Ohio in Canton, Youngstown, Lorain, and Leipsic. Because steel is a capital intensive, mechanized industry these plants are expected to produce about 630 jobs.[26] Many more construction jobs have been created in conjunction with these investment projects. As shale gas development expands overseas, these companies plan to expand their exports of energy industry related steel components.

The steel demand tied to shale gas expansion is mostly associated with drilling and infrastructure development. These activities are primarily front-loaded, taking place early in the development of the resource base. The industry must balance meeting these relatively short-run demands against long-run capacity expansion requirements to avoid excess capacity in the future.

While shale gas driven increases in demand have been important to the steel industry, the large supplies of low cost natural gas have been of less direct importance to the industry. The primary cost category in steel production that is likely to be affected by cheap natural gas is electricity; however, electricity costs account for only about 10% of total production costs.[27]

As a result, the cost reductions experienced by the steel industry are likely to be small. The primary cost in steel production is scrap steel which is unaffected by the cost of natural gas.

Transportation: Key to Energy Independence?

Concerns about U.S. energy security and energy independence revolve around the use of imported oil as the country is basically independent in all other forms of energy. The Obama Administration targeted a one-third reduction in oil imports from when the President took office or a 3.6 million barrel per day cut by the end of the decade.[28] While oil based products, gasoline and diesel fuel are the major fuels used in transportation, natural gas does not play a big role.[29] Although natural gas consumption by vehicles has grown by 38% from 2006 through 2011, it still represents less than 1% of U.S. natural gas consumption.[30] However, the low current and projected price of natural gas has the commercial truck industry, in particular, planning for ways to incorporate more natural gas into the transportation fuel mix. Even in EIA's most aggressive case for heavy duty vehicles running on natural gas, in 2035 natural gas comprises less than 9% of the highway vehicle fuel mix.[31]

Natural gas can become a greater part of the transportation fuel mix in a variety of ways: compressed natural gas (CNG), LNG, methanol, gas-to-liquids, fuel cells, and electricity. However, the primary near-term opportunities for natural gas in long-distance trucking are LNG and CNG. The use of natural gas as a transportation fuel in any form will require changes in the sector from consumers and suppliers. New vehicles will need to be developed on a large scale, in some cases requiring new technologies; new infrastructure will need to be deployed; and investments will need to be made by both consumers and industry.

Some inroads are being made in the transportation sector and additional efforts are being undertaken. Fleet trucks, including Waste Management, United Parcel Service, and AT&T among others, appear to be trying to take advantage of low natural gas prices.[32] Trucks can either be purchased new that run on natural gas, or conversion kits can be used to allow trucks that currently run on diesel fuel to use natural gas. Some retrofit conversions allow trucks to run on either LNG or diesel, allowing owners to take advantage of changing prices. The market for fuels for trucking is large. Approximately 3.2 million big-rig trucks use about 25 billion gallons or 1.7 million barrels per day, about 9% of U.S. oil consumption, of diesel fuel annually.[33]

Among the key factors in natural gas as a transportation fuel is the price spread between diesel fuel and natural gas. The cost of diesel fuel largely depends on the cost of crude oil. Historically, a barrel of oil cost about six times the cost of a unit of natural gas. As the price of oil has increased and the price of natural gas has fallen, a barrel of oil might cost over 30 times the cost of a unit of natural gas. This price spread has resulted in a price of diesel fuel twice as high as CNG on a diesel-gallon-equivalent basis.[34] The importance of the price spread between crude oil and natural gas demonstrates that natural gas can lose its competitive advantage because of a price rise, but it can also be lost if the price of crude oil falls. If a switch to natural gas-powered trucks increased the demand for new trucks many industries would expand production including the steel, tire, electronic, and vehicle manufacturing industries.

Long-haul trucking has made some progress to use natural gas, but concerns over the availability of refueling stations remain a hurdle. Shell announced a plan to add LNG pumps at 100 locations owned by Travel Centers of America for truck refueling.[35] The goal would be for LNG trucks to be able to travel across the entire country. Currently, city buses and local delivery trucks that return to a central facility where re-fueling is available are the primary consumers of natural gas in transportation. In addition, these users of natural gas tend to have short, or predicable, routes so that refueling can be reliably planned.

Residential and Commercial Consumption: A Regional Opportunity

Overall residential use of natural gas is not projected to increase as significantly as other sectors of the economy, and according to EIA estimates, may decline by 2035.[36] New infrastructure, both for consumers and industry, would be required to open new market areas to natural gas, and the increase in demand would have to warrant the investment. Nevertheless, there are some parts of the country where it may make sense.

As an example, in April 2011, New York City passed regulations to phase out the use of certain heating oils by 2030, and to replace them by cleaner-burning fuels, including natural gas.[37] However, new infrastructure will be required to bring additional volumes of natural gas, which will require local and federal approval. In 2010, New York State consumed almost 30 million barrels of distillate, which is one type of heating oil used for residential and commercial heating—almost half of the state's total distillate use.[38] Additionally, the states in New England consumed almost 50 million barrels of distillate in residential and commercial use. Expanding the natural gas infrastructure in these areas could help consumers take advantage of low natural gas prices, particularly given the region's proximity to the Marcellus Shale formation. Connecticut is an example of a state whose natural gas use is constrained by a lack of natural gas infrastructure.[39] The state is developing a plan to increase its natural gas consumption.

Increased sales of natural gas appliances, although a relatively small consumer group, could foster a change in perception of natural gas use. Besides the traditional appliances of stoves and hot water heaters, other household items like air conditioners could run on natural gas instead of electricity. Greater use of natural gas in this area could better integrate natural gas in the economy. Programs, like the Energy Star ratings on appliances, may need to be examined for how natural gas appliances are evaluated compared to other fuels.

Imports and Exports: A Reversal of Roles

Exports of natural gas have been on the rise, while imports have been declining. Increased exports, particularly as LNG, could provide a new demand center for U.S. supplies. Heading into the last decade, the United States was expected to be a growing importer of natural gas because domestic production was declining and demand was rising. EIA, in its 1999 Annual Energy Outlook, projected that net natural gas imports would grow between 1997 and 2020 from 12.9% of consumption to 15.5%, based on consumption growing faster than production.[40] To accommodate the potential increase in imports, five new LNG import terminals were built by industry in the latter half of the 2000s and some existing facilities were re-commissioned and expanded. The United States currently has LNG import capacity of almost 14 billion cubic feet per day (bcf/d) or over five trillion cubic feet (tcf) per year. However, higher domestic production—mainly from shale gas development—has made imports less necessary and they have been trending down over the last five years (see **Figure 11**). Import terminals are operating well below capacity. In its 2012 Annual Energy Outlook, EIA forecasts the United States becoming an overall net exporter of natural gas in 2022 and a net LNG exporter by 2016.

The abundance of new domestic natural gas supplies is shifting industry interest from building LNG import terminals to constructing LNG export terminals.[41] As of September 2012, 18 companies had applied for permits to construct liquefaction facilities at existing LNG import terminals or build new facilities, with a capacity of 27.4 bcf/d or 10.0 tcf per year.[42] Increased pipeline exports to Canada[43] and Mexico may also rise if those countries' domestic production continues to decline and their demand continues to increase. It is unclear if, when, and how much LNG exports will be allowed, but the effect on domestic prices is at the crux of the debate. Any exports would be an additional source of demand and would likely put upward pressure on prices.

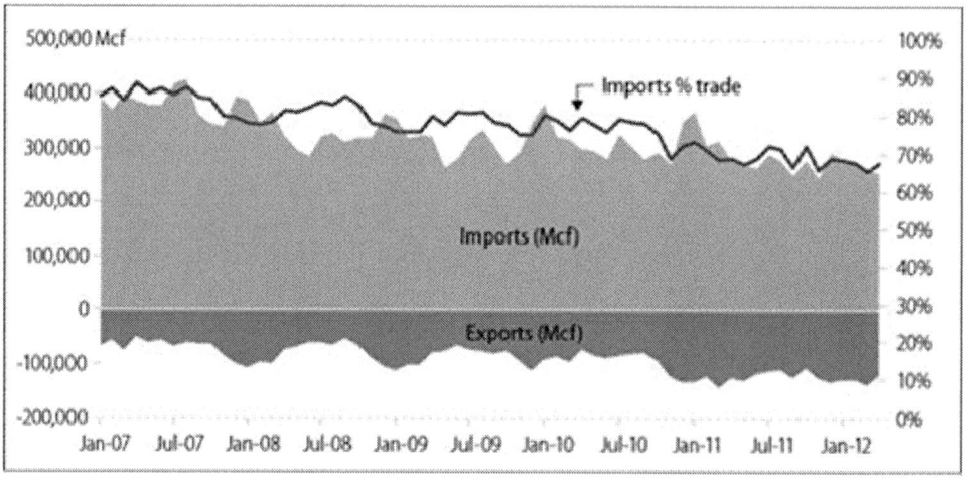

Source: U.S. Energy Information Administration.
Notes: Units = million cubic feet (Mcf).

Figure 11. U.S. Imports and Exports, 2007-2012.

Other Sectors That Could Gain

How natural gas ultimately filters through the economy and what other sectors may benefit is not clear at this point in time and there are many other industries that stand to benefit that have not been addressed in this report. The oil industry has probably been the biggest indirect beneficiary of shale gas development and low natural gas prices. Production companies have used the same techniques—directional drilling and hydraulic fracturing—to unlock tight oil formations and increase U.S. oil production. Also, the petroleum refining sector should gain as natural gas is used as a fuel and feedstock. The airline industry has tested synthetic aviation fuels derived from natural gas. In 2009, Qatar Airways flew the first commercial flight using a natural gas-based jet fuel.[44] The shipping industry is analyzing whether natural gas can be used as a bunker fuel. The paper and aluminum industries, which use large quantities of natural gas, may benefit. The aluminum industry has announced a couple of expansions and greenfield projects.

NATURAL GAS MARKETS: THE LIMITATIONS

Market forces in conjunction with a dynamic regulatory environment have the potential to raise the role natural gas contributes to the U.S. economy. Optimistic outlooks for rising supply and low prices have sparked interest in expanding the role of natural gas even further. Congress has expressed interest in a variety of the issues of this debate. Members and committees have introduced legislation, held hearings, and offered opinions and inquiries with government agencies.

Many producers and consumers of natural gas seem confident that abundant supplies of natural gas will be produced regardless of the risks still outstanding. Some companies on both the supply and demand side of natural gas have made, or have announced plans for large investments to confirm their positions. Other companies are waiting to see how various dimensions of the debate unfold. A key component of this discussion is how domestic prices will react under different market conditions. As demand rises so does the pressure on prices to rise, which signals more production is needed, or that consumption should be curtailed. The beneficial economic effects of expanded natural gas supplies are not pre-ordained. Market and regulatory factors could come into play, having the effect of reducing positive economic benefits for the United States.

The Global Market: A Role for the United States

The prospect of increasing U.S. LNG exports has become encumbered by concerns of higher domestic prices and increased volatility of prices. As required by the Energy Policy Act of 2005, the Department of Energy (DOE) must issue a permit to export natural gas to countries with which the United States does not have a free trade agreement (FTA).[45] DOE must also determine that export to non-FTA countries is in the public interest. As part of that determination DOE has undertaken a two-part study to examine the impact of LNG exports on domestic prices, which is expected to be completed in early 2013. It is not anticipated that any additional permits for export to non-FTA countries will be approved before part two is completed, and outside of South Korea, the FTA countries do not import a significant amount of LNG.

Exports of energy-related resources have not been a big part of U.S. trade so increasing them would likely improve the overall U.S. trade balance. Nevertheless, the rise in U.S. natural gas production has already benefitted the U.S. trade position by dramatically decreasing imports, the other component of the trade balance. Increasing U.S. LNG exports would also expand the role of the United States in international natural gas markets. Asian countries, in particular, and some European countries have called for more U.S. LNG exports. If all the proposed LNG exports came to fruition, which is unlikely, the United States could become the world's largest LNG exporter. The United States is already the largest producer and consumer of natural gas and has the most storage capacity of any country in the world. These factors, coupled with proposed construction of LNG export terminals on the East Coast, Gulf Coast, and West Coast, could propel the United States into the center of the global natural gas trade, benefiting the U.S. balance of trade.

Unlike some other countries, where the government controls natural resources, the private sector controls U.S. natural gas resources and is driven mostly by market forces. However, becoming a significant global supplier could enhance certain U.S. national security priorities, such as contributing to European natural gas supply diversification from Russia and providing a counter to the nascent gas cartel, the Gas Exporting Countries Forum, which includes Russia and Iran.[46] Other effects beyond national security could include market and price reforms, technology transfers, greater transparency, and environmental benefits.

The domestic situation may also have an effect on the rest of the world as major U.S. companies with large international portfolios, in particular, continue to buy into shale gas assets. These companies are becoming more natural gas oriented. Many have already started looking for opportunities to use the experience they have gained in the United States in other countries.

Additionally, many foreign companies have bought into U.S. shale gas areas to gain experience in producing these formations and eventually bring the technology back to their respective countries. Before the expansion of U.S. LNG exports occurs, a decision is likely to be made concerning whether energy security means relying primarily on captive domestic, or regional, supplies, or whether it means participating in a well-supplied, open world market. A large expansion of LNG exports means that the domestic demand for natural gas will increase. If that extra demand is not met with additional supply increases, prices are likely to rise. If prices rise, the economic advantage of U.S. LNG will diminish somewhat. Additionally, the natural gas consuming industries that forecasted low prices and, as a result, increased their capital investment will likely not experience returns that meet expectations. Consumers in all categories will likely face higher costs.

If exports of LNG are not permitted, or permitted at a low rate, most of the potential foreign trade benefits outlined in this report may not materialize. Prices could, nevertheless, rise if production is curbed.

Environmental Considerations[47]

With the advent of shale gas development, natural gas has been touted as a bridge fuel to a low carbon, renewable energy-based economy. Natural gas is cleaner burning than its hydrocarbon rivals—emitting less CO_2, particulate matter, sulfur dioxide, and nitrogen oxides, on average, than either coal or oil (see **Table 2**). But the use of new technologies and drilling practices, as well as the gas drilling boom they have engendered, has led to other environmental concerns and controversies.

These concerns centered initially on water quality issues, including the substantial use of water during hydraulic fracturing activities as well as the potential contamination of water by hydraulic fracturing chemicals and wastewater disposal. Concerns have since incorporated other issues, such as land use changes, potential for induced seismicity from produced water injection, infrastructure requirements, and emissions of air pollutants from extraction operations and transport.

The spectrum of concerns over hydraulic fracturing in unconventional reservoirs has led, in part, to various grassroots movements, some political opposition, and calls for regulatory actions and moratoria at the local, state, and federal levels.

Table 2. Air Pollution Emissions by Combusted Fuel Type Pounds per Billion Btu of Energy Input

Pollutant	Natural Gas	Oil	Coal
Carbon Dioxide	117,000	164,000	208,000
Carbon Monoxide	40	33	208
Nitrogen Oxides	92	448	457
Sulfur Dioxide	1	1,122	2,591
Particulates	7	84	2,744
Formaldehyde	0.750	0.220	0.221
Mercury	0.000	0.007	0.016

Source: Energy Information Administration (EIA), Office of Oil and Gas. Carbon Monoxide: derived from EIA, *Emissions of Greenhouse Gases in the United States 1997,* Table B1, p. 106. Other Pollutants: derived from Environmental Protection Agency, *Compilation of Air Pollutant Emission Factors, Vol. 1, 1998.*

Notes: No pre- or post-combustion removal of pollutants. Bituminous coal burned in a spreader stoker is compared with No. 6 fuel oil burned in an oil-fired utility boiler and natural gas burned in an uncontrolled residential gas burner. Conversion factors are: bituminous coal at 12,027 Btu per pound and 1.64% sulfur content; and No. 6 fuel oil at 6.287 million Btu per barrel and 1.03% sulfur content—derived from Energy Information Administration, Cost and Quality of Fuels for Electric Utility Plants (1996). Coal specifications can vary greatly and each variety has a different emissions profile.

Major oil and gas producing states have been reviewing and revising their oil and gas rules in response to technological changes in the industry, and specifically to advances in hydraulic fracturing and directional drilling. Although oil and gas exploration and production is regulated primarily at the state level, several regulatory developments are occurring at the federal level as well. The Environmental Protection Agency (EPA) has pursued initiatives related to hydraulic fracturing under three federal environmental statutes. In August 2012, EPA promulgated regulations under the authority of the Clean Air Act (CAA) that establish new air emissions standards for hydraulically fractured gas wells and other oil and gas production activities. The new rules entered into effect on October 15, 2012. Under the Safe Drinking Water Act (SDWA), EPA has issued draft Underground Injection Control (UIC) Program Guidance for Permitting Hydraulic Fracturing with Diesel Fuels.[48] This guidance is being developed in response to a provision of the Energy Policy Act of 2005 which revised the SDWA definition of underground injection to explicitly exclude the underground injection of fluids (other than diesel fuels) used in hydraulic fracturing operations. Final guidance is expected in Spring 2013. Under the Clean Water Act (CWA), EPA is developing regulations governing the discharge of wastewater produced in natural gas extraction from shale formations and coal beds. The regulations will be established under EPA's effluent guidelines program, which sets national standards for the discharge of pollutants from industrial activities directly to surface waters and municipal wastewater treatment facilities. EPA plans to propose regulations for the coalbed methane effluents in 2013, and for the shale gas extraction regarding wastewater pollutants in 2014.[49]

In May 2012, the Bureau of Land Management (BLM), Department of the Interior, proposed revisions to its oil and gas development rules in response to the increased use of hydraulic fracturing on federal and Indian lands. The proposal would require public disclosure of chemicals used during hydraulic fracturing, tighten requirements related to well-bore integrity, and add requirements for managing water used and produced in hydraulic fracturing operations. The public comment period closed September 10, 2012, and the BLM is now reviewing comments. A tighter regulatory environment for natural gas exploration and production, if it raises costs significantly, would likely result in slower supply growth, and in the extreme could reduce some of the economic benefits described in this report. If demand increases as the result of expectations of rising supply and low prices, but then regulation slows supply growth, a price spike is the likely result. A price spike could reduce market confidence and set back the use of natural gas, especially in the industrial sector. Some companies in the industrial demand sector, as mentioned earlier, have made significant investments predicated on the supply being available and prices remaining low. However, in the past several years, a number of states have adopted stricter production regulations, yet production has continued to grow in those states (e.g., Colorado, North Dakota, Ohio, Pennsylvania, and Wyoming). In contrast, New York State essentially has a moratorium on high-volume hydraulic fracturing while regulations are revised. Overall, it is difficult to generalize regarding the potential effect of regulations. For example, EPA projects that the capture of natural gas required under its new air rules could have a net positive benefit for natural gas producers because of the value of the gas that could be saved and sold.

Price Levels and Volatility: Not a Certainty

The ability of U.S. industry to be positively affected by the evolving position of the United States with respect to natural gas supply depends not only on the low current price, but the expectation that prices will remain low into the future. The expectation of low and non-volatile prices for natural gas is not easy to relate to the historical price data shown in **Figure 3**. In the recent era of high prices, the assumption of many analysts was that prices were likely to remain high and increasing well into the future. This pessimistic conclusion was reached by considering the upward trend of consumption coupled with a declining resource base. Earlier, it was believed that natural gas was so abundant it could be treated as essentially a good with little value. These earlier failed expectations suggest that what appears to be a clear future path for natural gas markets can quickly reverse itself.

> Considerable uncertainty exists regarding the size of the economically recoverable U.S. shale gas resource base and the cost of producing those resources. Across four shale gas resource scenarios from the Annual Energy Outlook 2012 (AEO2012), natural gas prices vary by about $4 per million British thermal units (MMBtu) in 2035, demonstrating the significant impact that shale gas resource uncertainty has in determining future natural gas prices. This uncertainty exists primarily because shale gas wells exhibit a wide variation in their initial production rate, rate of decline, and estimated ultimate recovery per well (or EUR, which is the expected cumulative production over the life of a well).[50]

Natural gas demand has not exhibited dramatic changes to correspond with its price volatility. For many of the final consuming sectors, households, commercial uses, and electric power generation the demand has been relatively price inelastic, insensitive to changes in price. This is because of the nature of natural gas use, primarily heating in households and commercial uses. Only in the industrial demand sector is there significant price elasticity as shown by the reductions in demand and plant closures associated with the last period of high prices.[51] The inelasticity of natural gas demand also means that relatively small variations in the quantity available can have surprisingly large effects on price. This is because, with inelastic demand, any shortage or surplus of supply requires a disproportionate variation in price to ration people out of or bring consumers into the market. Price is also important to the potential expansion of the supply of natural gas. If prices fall, and remain, too low to provide an adequate return for producers, exploration and development will stagnate and future supply projections will prove to be optimistic. However, if the price of natural gas rises, and returns to producers rise, demand might be discouraged. The market must price natural gas in such a way that consumers are drawn to use more natural gas and that producers are encouraged to produce more natural gas. Consumers must also be incentivized to undertake the expense of changing fuels. Maintaining the status quo is a powerful force and overcoming it requires motivation beyond short-term gains because of prices. The market has not generated such razor-edged prices in the past. Not only the price of natural gas will determine the state of the market. The price of oil and coal are also likely to play an important role. If the price gap between crude oil narrows, natural gas will become a less attractive source of energy than oil. The reason is likely to be habit and infrastructure. Consumers are used to using oil-based products and the infrastructure to supply them is well established. If the incentive of cost savings disappears, it is likely that the mass of consumers will continue using established products. Similarly, the cost of a new supply infrastructure is likely to be avoided if the anticipated demand growth of natural gas becomes more uncertain. For coal, electric power plants may be idled for some time while fuel prices give natural gas an advantage. But, if natural gas prices rise because of increased demand, coal may regain market share.

Demand Competition

Enthusiasm over the state of natural gas markets has led to plans of increased consumption in all demand categories. However, demand expansion differs among the EIA's demand segments. One characteristic that is common is that expanded demand only occurs after a lag. Low prices increase interest in expanding consumption, but then investments must take place before growth in consumption is realized. Once the investments are in place, which can take years, in many parts of the market it is hard to reverse them and go back to previously used fuels. Expansion of household and commercial demand usually means natural gas for use in space heating. Conversion from other fuels is minimal, because of inadequate supply infrastructure, so the main source of new demand is new construction. Once a heating system in new construction is in place it is unlikely to be replaced by a different fuel. As a result, demand increases from these sectors are likely to be incremental and long-term. Similarly, while the demand for natural gas in transportation is likely to exhibit sharp percentage increases relative to its small base, its overall effect on demand is likely to be incremental and permanent. Once vehicle conversions or vehicles with dedicated power plants are on the road and the natural gas supply infrastructure is in place, demand growth is likely to continue. After buying a home, a car is probably the next largest investment for most households and not likely to be changed because of short-term price changes. Electric power generation and industrial use of natural gas are different. Much more rapid demand expansion can occur in these segments. Each consumer uses large quantities of gas and total demand can increase sharply. If the export of LNG expands, it would exhibit similar characteristics. Industrial demand is also likely to be price sensitive. Industrial demand has contracted in periods of high prices.

The combination of trend growth generated by household and commercial demand coupled with periodic large increases in demand associated with investment by power generators and industrial users could, if all planned expansions come to fruition, lead to a demand level which is too large relative to supply growth after the period of low prices. The result would likely be upward price volatility which could destabilize the market and reduce expected economic benefits.

Other Factors

Other legislative or regulatory policies could impact the natural gas market. For example, if legislation or regulation required tight carbon emission limits for electric power generators and the price of natural gas was low, a large increase in demand might be expected. Increases in demand from any source tend to raise prices. However, the price increases would tend to lead to increased development of natural gas resources and increased supply.

If offshore drilling opportunities were expanded, as oil and natural gas were discovered, domestic supplies would increase, causing the price of natural gas to fall. Falling prices would benefit consumers in the short-term, but would likely reduce the development of on-shore shale resources in the longer term, depending upon which sources had lower costs.

These two examples show that unintended consequences of policy decisions could affect natural gas markets. The results could upset the calculation of net economic benefits that the United States might expect to experience from development of the expanded shale gas resource base.

Conclusion

Given existing data, most indications point to the changes in the natural gas industry as positive to the overall U.S. economy. Industries that use natural gas as an input have seen prices fall. Producers have expanded the U.S. resource base tremendously, including for oil. Low prices have forced producers to be innovative and drive down production costs. Similarly, environmental concerns are prompting companies to be more proactive in addressing these issues.

Also responding to environmental concerns, a number of major natural gas and oil producing states have revised their rules, which has not appeared to have inhibited production in those states.

The next five years will be telling as proposed projects will either come to fruition, be postponed, or be cancelled. Investment decisions are being made given existing realities and future perceptions of the market. Many of the environmental concerns may be addressed by industry, federal, state, and local governments, and the market.

Production efficiencies and extraction improvements, particularly in regards to water use, are likely to be made in this time period. Congress may act on a variety of issues that would affect natural gas. Legislation has been introduced that could increase the supply of natural gas by opening up more federal lands and offshore areas to exploration, by providing direct incentives for natural gas production, and by limiting federal regulations on federal lands. Other legislation may inhibit natural gas production by eliminating incentives, increasing regulatory requirements on companies, or limiting access to potential areas of development. Legislation has also been introduced that could increase demand for natural gas by vehicles and in aviation.

Although some Members have expressed support for LNG exports, legislation has been introduced that would delay LNG exports, thereby limiting a potential source of demand. Still other legislation has been introduced that would promote natural gas infrastructure in some areas and other legislation has been introduced to limit the environmental effects of natural gas production. Similarly, rules and regulations are being proposed at different federal agencies, for example the Departments of Energy and Interior, and the Environmental Protection Agency, which may require or elicit congressional input.

APPENDIX A

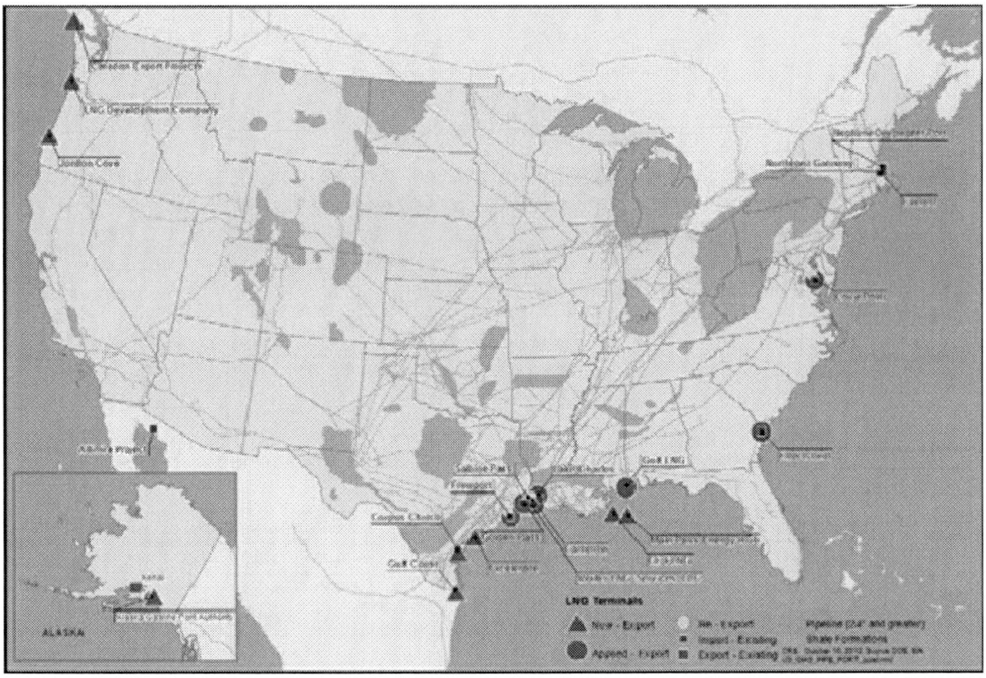

Source: Compiled by CRS using data from the Department of Energy and the U.S. Energy Information Administration.

Figure A-1. U.S. Natural Gas Infrastructure and Shale Resources.

APPENDIX B

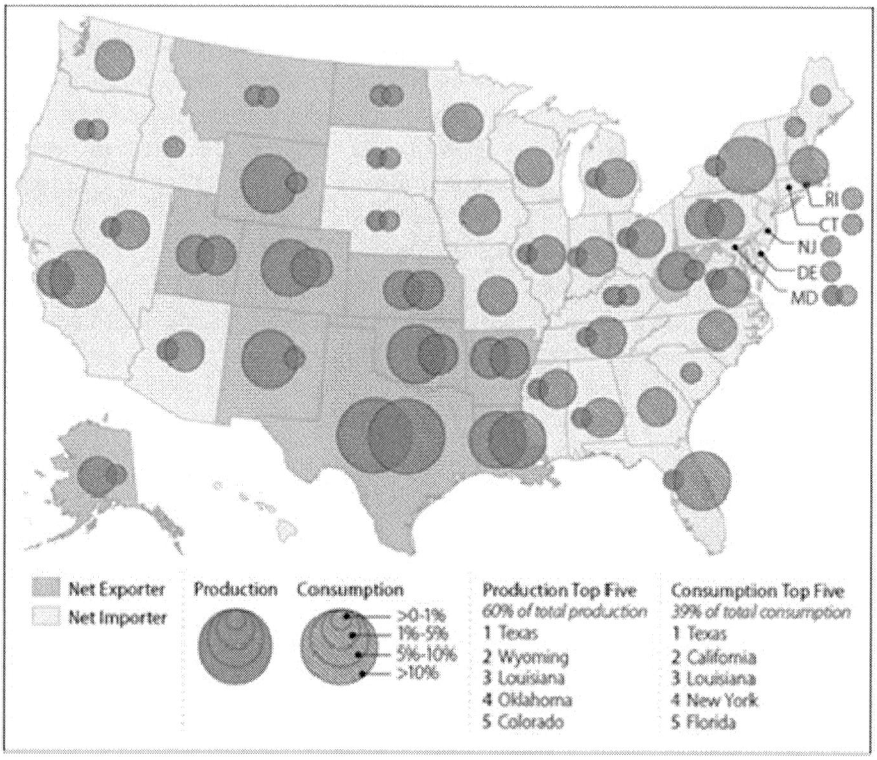

Source: U.S. Energy Information Administration, http://www.eia.gov/dnav/ng/ng_prod_sum_dcu_NUS_a.htm and http://www.eia.gov/dnav/ng/ng_cons_sum_dcu_nus_a.htm.

Notes: Production is dry gas production.

Figure B-1. State Production and Consumption, 2011, % of total U.S. production and consumption.

APPENDIX C

Table C-1. Energy Conversions and Comparisons

Fuel	Unit	Btu Equivalent
Natural Gas	1 cubic foot	1,032 Btu
Gasoline	1 gallon	124,238 Btu
Diesel	1 gallon	138,690 Btu
Coal	1 short ton	19,858,000 Btu
Electricity	1 kilowatthour	3,412 Btu

Source: U.S. Energy Information Administration, http://www.eia.gov/energyexplained/ index.cfm?page= about_energy_units.

Notes: Btu = British thermal unit.

End Notes

[1] Reserves is an industry term to define the likelihood that natural gas resources can be produced using current technology and at today's prices according to the Society of Petroleum Engineers and the World Petroleum Congresses definition.

[2] Many producing countries subsidize natural gas consumption, so their consumers do not pay a market price.

[3] 2010 is the latest data is available for this metric.

[4] Spot price data is from the Energy Information Administration, available at http://www.eia.gov/dnav/ng/ng_pri_fut_s1_d.htm.

[5] For additional information on the shift to tight oil see CRS Report R42032, *The Bakken Formation: Leading Unconventional Oil Development*, by Michael Ratner et al.

[6] U.S. Energy Information Administration, *Natural Gas Gross Withdrawals and Production*, database, 2010, http://www.eia.gov/dnav/ng/ng_prod_sum_dcu_snd_a.htm.

[7] For additional information on the Marcellus Shale Formation, see CRS Report R42333, *Marcellus Shale Gas: Development Potential and Water Management Issues and Laws*, by Mary Tiemann et al.

[8] For additional information on the role of water in natural gas production, see CRS Report R41507, *Energy's Water Demand: Trends, Vulnerabilities, and Management*, by Nicole T. Carter and CRS Report R42333, *Marcellus Shale Gas: Development Potential and Water Management Issues and Laws*, by Mary Tiemann et al.

[9] Department of Energy, Office of Fossil Energy and the National Energy Technology Laboratory, *Modern Shale Gas Development in the United States: A Primer*, DE-FG26-04NT15455, April 2009, pp. 58-59, http://fossil.energy.gov/ programs/oilgas/publications/ naturalgas_general/Shale_Gas_Primer_2009.pdf.

[10] U.S. Geological Survey, *Summary of Water Use in the United States, 2000*, http://ga.water.usgs.gov/edu/wateruse2000.html.

[11] U.S. Energy Information Administration, *Net Generation by Energy Source: Total (All Sectors), 2002-June 2012*, July 2012, http://www.eia.gov/electricity/ monthly/epm_table_ grapher.cfm?t=epmt_1_1.

[12] U.S. Energy Information Administration, *U.S. Energy-Related CO2 Emissions in Early 2012 Lowest Since 1992*, August 1, 2012, http://www.eia.gov/todayinenergy/detail.cfm?id=7350.

[13] There are three major regional natural gas markets; North America, Europe, and Asia. There is no recognized world price of natural gas comparable to that of crude oil. Historically, natural gas markets have been characterized by longterm contracts indexed to the price of oil, rather than demand and supply conditions. There is limited international trade in natural gas, likely insufficient to cause prices to converge in the regional markets.

[14] Clifford Krauss and Eric Lipton, "After the Natural Gas Boom," *New York Times*, Sunday Business, October 21, 2012.

[15] This section draws upon IHS Global Insight, "The Economic and Employment Contributions of Shale Gas in the United States," December 2011.

[16] As noted in the report, natural gas is already displacing coal in electric power generation, and retirement of older coal-fired generation has been announced. Natural gas emits less pollutants than coal when combusted, see Table 2, on a fuel comparison basis. When factoring in the efficiency of natural gas power plants compared with the coal facilities being retired, the emissions improvements are even greater.

[17] U.S. Energy Information Administration, *Annual Energy Outlook 2012*, June 25, 2012, http://www.eia.gov/forecasts/ aeo/sector_electric_power_all.cfm#powergen.

[18] For additional information on natural gas displacing coal in electric power generation, see CRS Report R41027, *Displacing Coal with Generation from Existing Natural Gas-Fired Power Plants*, by Stan Mark Kaplan. Please note that Mr. Kaplan is no longer with CRS, and questions can be referred to Richard Campbell with CRS.

[19] Shauk, Zain, "Cheap Natural Gas Feeds Chemical Industry Boom," *Houston Chronicle,* April 19, 2012.

[20] "Shale Gas and New Petrochemicals Investment: Benefits for the Economy, Jobs, and U.S. Manufacturing," Economics and Statistics, American Chemistry Council, March 2011.

[21] There are some U.S. petrochemicals companies that use naphtha as a feedstock in their production, and many are believed to be shifting to NGLs. It does not appear that any possible negative consequences were factored into the American Chemistry Council study.

[22] Reuters, "U.S. Shale Gas Boom Brings Bad News for Europe's Oil Refiners," July 26, 2012.

[23] Simon Constable, "How Corn Is Feeding Fertilizer," available at http://barrons.com/article/SB50001424053111903835404577347881696361296.html, April 21, 2012.

[24] John W. Miller, "Steel Finds a Sweet Spot in the Shale," *Wall Street Journal*, March 26, 2012.

[25] Keith Schneider, "As Demand Rises, Ohio's Steel Mills Shake Off the Rust and Expand," *New York Times*, April 24, 2012.

[26] Ibid.

[27] IHS Global Insight, "The Economic and Employment Contributions of Shale Gas in the United States," December 2011.

[28] The White House, *Remarks by the President on America's Energy Security*, Washington, DC, March 30, 2011, http://www.whitehouse.gov/the-press-office/2011/03/30/remarks-president-americas-energy-security.

[29] For additional information on natural gas vehicles, see CRS Report RS22971, *Natural Gas Passenger Vehicles: Availability, Cost, and Performance*, by Brent D. Yacobucci.

[30] U.S. Energy Information Administration, http://www.eia.gov/dnav/ng/ng_cons_sum_dcu_nus_ a.htm.

[31] Energy Information Administration, *Annual Energy Outlook*, DOE/EIA-0383(2012), June 24, 2012, http://www.eia.gov/forecasts/aeo/index.cfm.

[32] Rebecca Smith, *Wall Street Journal*, online edition, "Will Truckers Ditch Diesel?", Business Section, May 23, 2012.

[33] Ibid.

[34] Ibid.

[35] Christopher Helman, "Shell Investing $300M To Fuel LNG-Powered Trucks," *Forbes*, June 13, 2012.

[36] U.S. Energy Information Administration, *Annual Energy Outlook 2012*, Washington, DC, June 2012, http://www.eia.gov/oiaf/aeo/tablebrowser/#release=AEO2012&subject=2-AEO2012 &table=2-AEO2012&region=1- 0&cases=ref2012-d020112c.

[37] Bill Holland, "Bloomberg Doubles Down on Natural Gas for New York City," *Platts*, August 28, 2012, online edition.

[38] U.S. Energy Information Administration, *Sales of Distillate Fuel Oil by End Use*, http://www.eia.gov/dnav/pet/pet_cons_821dst_dcu_nus_a.htm.

[39] John Kemp, "Connecticut Contemplates Connecting More To Gas Network," *ThompsonReuters*, October 19, 2012, online edition.

[40] U.S. Energy Information Administration (EIA), *Annual Energy Outlook 1999 with Projections to 2020*, DOE/EIA-0383(99), Washington, DC, December 1998, p. 71, http://www.eia.gov/ oiaf/archive/aeo99/pdf/0383(99).pdf.

[41] For additional information on U.S. natural gas exports, see CRS Report R42074, *U.S. Natural Gas Exports: New Opportunities, Uncertain Outcomes*, by Michael Ratner, Paul W. Parfomak, and Linda Luther.

[42] Department of Energy, Office of Fossil Energy, *Summary of LNG Export Applications*, Washington, DC, September 21, 2012, http://fossil.energy.gov/programs/gasregulation/ reports/Long_Term_LNG_ Export_ Table_08_17_12.pdf.

[43] For additional information on the U.S.-Canada energy relationship see CRS Report R41875, *The U.S.-Canada Energy Relationship: Joined at the Well*, by Paul W. Parfomak and Michael Ratner.

[44] Adam Schreck, "Qatar First to Fly with Jet Fuel Based on Natural Gas," *Associated Press*, October 13, 2009, online edition.

[45] Free Trade Agreement countries that require national treatment include Australia, Bahrain, Canada, Chile, Costa Rica, Dominican Republic, El Salvador, Guatemala, Honduras, Jordan, Mexico, Morocco, Nicaragua, Oman, Peru, Singapore, and South Korea.

[46] For additional information on European energy security, see CRS Report R42405, *Europe's Energy Security: Options and Challenges to Natural Gas Supply Diversification*, coordinated by Michael Ratner.

[47] Research by Mary Tiemann and Richard Lattanzio contributed to this section.

[48] For additional information on the role of the Safe Drinking Water Act, see CRS Report R41760, *Hydraulic Fracturing and Safe Drinking Water Act Issues*, by Mary Tiemann and Adam Vann.

[49] Environmental Protection Agency, http://yosemite.epa.gov/opei/RuleGate.nsf/.

[50] U.S. Energy Information Administration, *Projected natural gas prices depend on shale gas resource economics*, August 27, 2012, http://www.eia.gov/todayinenergy/detail.cfm? id=7710#.

[51] See CRS Report R41628, *Industrial Demand and the Changing Natural Gas Market*, by Robert Pirog.

INDEX

D

G

H

I

J

L

M

N

O

P

Q

R

S